Mathematical Modeling the Life Sciences

The purpose of this unique textbook is to bridge the gap between the need for numerical solutions to modeling techniques through computer simulations to develop skill in employing sensitivity analysis to biological and life sciences applications.

The underpinning mathematics is minimized. The focus is on the consequences, implementation, and application. Historical context motivates the models. An understanding of the earliest models provides insight into more complicated ones.

While the text avoids getting mired in the details of numerical analysis, it demonstrates how to use numerical methods and provides core codes that can be readily altered to fit a variety of situations.

Numerical scripts in both Python and MATLAB® are included. Python is compiled in Jupyter Notebook to aid classroom use. Additionally, codes are organized and available online.

One of the most important skills requiring the use of computer simulations is sensitivity analysis. Sensitivity analysis is increasingly used in biomathematics. There are numerous pitfalls to using sensitivity analysis and therefore a need for exposure to worked examples in order to successfully transfer their use from mathematicians to biologists.

The interconnections between mathematics and the life sciences have an extensive history. This book offers a new approach to using mathematics to model applications using computers, to employ numerical methods, and takes students a step further into the realm of sensitivity analysis. With some guidance and practice, the reader will have a new and incredibly powerful tool to use.

Nicholas G. Cogan is Professor of Mathematics at Florida State University. He began studying mathematical biology in undergraduate school and received his Ph.D. from the University of Utah under James P. Keener. He routinely works with microbiologists, environmental engineers, clinicians, and other scientists outside of mathematics. He has taught for twenty years at the undergraduate level and his research focuses on mathematical modeling in the life sciences. He is the author of over fifty articles using mathematics with biology.

Textbooks in Mathematics

Series editors:

Al Boggess, Kenneth H. Rosen

An Introduction to Complex Analysis and the Laplace Transform
Vladimir Eiderman

An Invitation to Abstract Algebra
Steven J. Rosenberg

Numerical Analysis and Scientific Computation
Jeffery J. Leader

Introduction to Linear Algebra
Computation, Application and Theory
Mark J. DeBonis

The Elements of Advanced Mathematics, Fifth Edition
Steven G. Krantz

Differential Equations
Theory, Technique, and Practice, Third Edition
Steven G. Krantz

Real Analysis and Foundations, Fifth Edition
Steven G. Krantz

Geometry and Its Applications, Third Edition
Walter J. Meyer

Transition to Advanced Mathematics
Danilo R. Diedrichs and Stephen Lovett

Modeling Change and Uncertainty
Machine Learning and Other Techniques
William P. Fox and Robert E. Burks

Abstract Algebra
A First Course, Second Edition
Stephen Lovett

Multiplicative Differential Calculus
Svetlin Georgiev, Khaled Zennir

Applied Differential Equations
The Primary Course
Vladimir A. Dobrushkin

Introduction to Computational Mathematics: An Outline
William C. Bauldry

Mathematical Modeling the Life Sciences
Numerical Recipes in Python and MATLAB®
N. G. Cogan

https://www.routledge.com/Textbooks-in-Mathematics/book-series/CANDHTEXBOOMTH

Mathematical Modeling the Life Sciences

Numerical Recipes in Python and MATLAB®

N. G. Cogan

CRC Press
Taylor & Francis Group
Boca Raton London New York

CRC Press is an imprint of the
Taylor & Francis Group, an **informa** business

A CHAPMAN & HALL BOOK

First edition published 2023
by CRC Press
6000 Broken Sound Parkway NW, Suite 300, Boca Raton, FL 33487-2742

and by CRC Press
4 Park Square, Milton Park, Abingdon, Oxon, OX14 4RN

CRC Press is an imprint of Taylor & Francis Group, LLC

ISBN: 9780367554934 (hbk)
ISBN: 9781032328263 (pbk)
ISBN: 9781003316930 (ebk)

DOI: 10.1201/9781003316930

Publisher's note: This book has been prepared from camera-ready copy provided by the authors.

I am deeply grateful to my mentors. Clyde Martin invited me into the field of mathematical biology when I was an unruly undergraduate. Jim Keener taught me how to look at problems without fear. Lisa Fauci taught me how to think deeply about a problem, care about the details, and how to mentor. Yousuff Hussaini taught me to think about the big picture and look for connections across disciplines and topics. I have also been blessed with a wonderful group of colleagues and students at Florida State University and cannot be grateful enough for this community.

<u>*However*</u>

There is no doubt that my family is the most important thing in my life. My wife has supported me, pushed me, and loved me. My children have challenged me, inspired me, and encouraged me. I could not have done this without them.

Contents

Foreword

This book was originally conceived as a book by a mathematician but intended for biologists – or at least readers whose primary interest is in biological insights. This seems an impossible task since there are inherent differences in the two disciplines. Biology is primarily an observational science while mathematics is primarily a constructive science. However, mathematical biology as a subdiscipline of both mathematics and biology has moved from a small community into an incredibly active and diverse field. This is because both mathematics and biology benefit from the close communication.

But this leads us to ask what pieces of information or skills are required? How much mathematics does a biologist need to really know? How much biology does a mathematician really need to know? Our answer is "Enough to do what is necessary". With that as our philosophy, we have tried to minimize the underpinning of the mathematics and focus on the consequences and implementation/application. It really is important to understand where some of the analysis comes from – especially if you ever want to do anything that is even slightly beyond the scope of this book. But, we have also provided numerical codes that are reasonably generic. In our view, most of dangers of numerical packages is pushing them beyond what they are intended for, so we have tried to include some problems that show you danger signs to look out for. Similarly, we do not claim to be inclusive or complete regarding any of the biological topics. We have tried to motivate the models from a historical context since an understanding of the earliest models and biological questions provide insight into the more complicated models in the current literature.

The main important theme in this book that differentiates it from other mathematical biology texts, is sensitivity analysis. This was introduced more than 100 years ago and it has become more widely used. But very few people know where sensitivity measures come from or what methods can be used. Even fewer people have access to robust methods that are simple to implement for a variety of problems. Most packages are either not flexible enough or so packed that it is very daunting for a non-expert to think about using them. Hopefully with some guidance and practice, the reader will have a new and incredibly powerful tool to use.

The book is aimed at undergraduates, but then almost all the books I used in graduate school claimed the same thing. I have used the material presented here with both graduates and undergraduates. I feel that the material is too cursory for a decent graduate course in mathematics – there are many other mathematical biology textbooks that are more complete. But, even in a graduate curriculum in mathematics, there is a need for practice in implementation.

Our guess is that the notation is a bit daunting for a typical undergraduate biology major. However with a single course in calculus and a bit of priming in differential equations, the material is accessible. In fact, this is one of the aims of the presentation – we believe that a practicing biologist should have a course in calculus and really needs some understanding of differential equations. But not what is typically taught in a mathematics department. Most real biological processes avoid the issues that are the heart of the mathematics of differential equations. Dynamical systems (i.e. differential equations where time is the underlying independent variable) are reasonably well behaved unless the defining terms are complicated (or more specifically if there are discontinuities, or certain nonlinearities). But these are typically not found in biologically motivated problems and are definitely not the place to start! On the other hand, almost all interesting behaviors in the real world are nonlinear. A first course in differential equations often spends the majority of the time on linear equations. So how do we bridge this gap between the relatively simple and the relatively useful? Our main tools are numerical. Along with demonstrating sensitivity analysis, we show many examples of manipulating differential equations and numerical methods. With this as a foundation, there are many very practical problems that are within reach.

If a student has no exposure to differential equations, some study in Chapter 2 will be required. This introduces the main tools that are needed: Steady-states, stability, qualitative analysis, and time-course plots. One of the most important goals in the book is to allow, encourage, and in some sense require that the students can use numerical methods to calculate much of this. We have provided scripts that are relatively generic and can produce all of these. Of course, there is no such thing as a truly black-box script and the user *has* to be responsible for understanding that numerical methods can miss steady-states, produce erroneous estimates and are not able to check the users signs! The rule of GIGO (Garbage In, Garbage Out) is not just a rule of thumb but a governing principle. But the methods described and the example models provides a working template that will be useful to a wide variety of scientists.

The organization of the book is relatively linear in the topic of sensitivity analysis. The first few chapters deal with one-at-a-time methods and those that are primarily visual or geometric. For these, we have included numerical scripts that show the logic and implementation.

The purpose of these files is to show the flow of information and they are not designed for efficiency. In the later chapters, we introduce more sophisticated methods (e.g. Screening methods, PRCC, Sobol'). We have chosen to show how to use other packages for these rather than show the native code. Partly, the implementation requires information that is far beyond the scope of this class to understand – for example almost all implementations of Sobol' indices use Monte Carlo integration. A second reason for showing this is to show that it is possible to use very sophisticated methods as long as the student has sufficient background. PRCC, Sobol', and screening methods are used in high-level, cutting-edge research – and it is within reach of the students.

On the other hand, the biological topics are completely interchangeable. We have attempted to use examples that are relevant for sensitivity methods. For example, it is not useful to use a screening method for a small number of parameters. But an instructor should be able to pull biological material from any chapter. The numerical scripts are reasonably generic and after a few weeks of exposure I have found students are able to manipulate the codes very easily.

There is some required background, mostly from the mathematical side. A sequence in calculus and some exposure to differential equations is really all that is needed. The first three chapters describe the philosophy of modeling which is typically not well articulated in textbooks. This is followed by some mathematical requirements and advice on numerical practices. When I have taught from this material, this comprises about two weeks of a semester – then the interesting material is accessible.

We use both MATLAB® and Python for the numerical portions. They both have strengths and weaknesses but it is very good to see that they are quite similar. Being able to implement, run and read code in both is quite useful. We will eventually include R and Julia since these are becoming wide-spread. In my opinion, there is little reason to learn uninterpreted languages like Fortran and C for most people who focus on modeling. The models change too quickly for any payoff in terms of efficiency.

Numerical methods can be complemented by symbolic methods like Mathematica and Maple. In my opinion these packages are not well suited for these sorts of methods – however, it is useful to be aware of them.

Finally, I hope that the material is interesting – I am fascinated by the historical setting for both the science and the mathematics. The beginning of each chapter covers some of the relevant history. I hope it interests the reader enough to dig deeper into where these topics came from.

N. G. Cogan
Tallahassee, Fl

1

Introduction

Modeling is a central concept in mathematical biology since there are not fundamental "laws" to rely on. Therefore it is important to have an idea of a model in mind. By definition, models are approximations of reality. Good models are capable of providing insight into biological hypotheses and built to be refined and adjusted.

All models are wrong, but some are useful.

Attributed to George E. P. Box

Models should be as simple as possible, but no simpler.

Adaptation of Albert Einstein

1.1 What Is a Model?

Models are the translation of a real world problem into a statement or question in mathematical language. Models connect the lab scientist, clinician, or practitioner to the mathematician/quantitative scientist. Bridging this gap requires an understanding of both languages, as well as an appreciation of the gap between them.

Like translations, there can be good and bad translations – those that are accurate, those that capture the spirit of the primary language and those that open up insights. So what makes a good model? This depends a lot of the context. Some qualities that make good models

DOI: 10.1201/9781003316930-1

include accuracy (how true to the science is the model), flexibility (if you change the experiment slightly, do you need a brand new model?), simplicity (is the model tractable?), practicality (does it include things that are not measurable?).

1.2   Projectile Motion

There are many relationships that are really models. For example, $D = rt$ models the how the distance travels depends on the rate and time. The ideal gas law relates pressure, volume, and temperature, $PV = nRT$. Einstein derived a model describing the relationship between energy and mass, $e = Mc^2$. All of these are models that provide insight into different processes. There are distinctions between these models. It is hard to imagine how $D = rt$ could not be an accurate model (although there is an underlying assumption that the rate is constant) while the ideal gas law becomes less and less accurate when the gas is close to a phase transition, say near the condensation point. The last relationship depends on assumptions underlying much of modern physics. So models can come in a range of subjective interpretation.

We will detail one model that is familiar to any student in calculus – projectile motion. We assume that an object is fired directly upward from a certain height above ground. The object goes up, stops moving, reverses direction and returns down. A description of the height of the object as a function of time is $y(t) = -\frac{1}{2}gt^2 + v_0 t + y_0$. Gravitational acceleration is denoted g, the initial velocity and positions are v_0 and y_0, respectively. There is a choice in frame of reference where we are assuming that gravity only acts in one direction (the object is dropped or thrown vertically with no horizontal displacement) and that gravity causes acceleration in the "negative" direction. If the object starts with a positive height, y_0 and is released the height will decrease.

This is often presented as "projectile motion" and as though it is always true. We could think of this function as the model – we can make predictions about the time it takes to return to $y = 0$ or what is the maximum height the object reaches. However, viewing this relationship that way obscures a broader modeling method – How do we adjust this for multiple objects? Is this accurate in all velocity regimes? What about projectile motion in different environments?

In calculus you might have seen this as a question about motion under constant acceleration. Acceleration is the rate of change of velocity, $v(t)$ and Newton argued that acceleration due to gravity is constant so that,

$$\frac{dv}{dt} = -g.$$

Since g is a constant, we can integrate both sides,

$$v(t) = -gt + c_0.$$

Now we know how the velocity changes over time and velocity is the rate of change of position,

$$\frac{dy}{dt} = -gt + c_0.$$

Since g and c_0 are constant, we can integrate both sides,

$$y(t) = -\frac{1}{2}gt^2 + c_0 t + c_1. \tag{1.1}$$

You can almost see the correspondence between $y = -\frac{1}{2}gt^2 + v_0 t + y_0$ and $y(t) = -\frac{1}{2}gt^2 + c_0 t + c_1$, but we have not yet determined the integration constants. We have to provide some information to do this. If we denote the initial position – the position at time 0 as $y(0) = y_0$, and plug this into 1.1, we find that $c_1 = y_0$.

So what is c_0? One way to determine it is if we define the initial velocity, $v(0) = v_0$. Then the velocity (the derivative of the position, $y(t)$) is,

$$\frac{dy(t)}{dt} = -gt + c_0. \tag{1.2}$$

Evaluating this at $t = 0$, we find that $c_0 = v_0$.

So the "model" $y = -\frac{1}{2}gt^2 + v_0 t + y_0$ is really the solution to the model,

$$\frac{d^2y}{dt^2} = -g. \tag{1.3}$$

This is a differential equation, and if we prescribe information $y(0) = y_0$ and $\frac{dy}{dt}(0) = v_0$, we can solve this completely.

But where did the model come from? Part of the point of this book is to learn how to invent these sorts of models, so we won't go into too much detail here. However, this is a nice example that has roots to Newton and the time that calculus was created. One of Newton's most valuable contribution is the idea of Laws – that are general principles that lead to mathematical statements. One of his laws of motion (the second) states that $f = ma$. That is, there is a balance between forces applied on an object and the acceleration of the objects mass. So Equation 1.3 is really a translation of Newton's second law (with a bit of algebra),

$$
\begin{aligned}
ma &= f, \\
\frac{d^2 y(t)}{dt^2} &= \frac{f}{m}, \\
&= -g.
\end{aligned}
$$

This is a model of the position of an object under constant acceleration. From a modeling standpoint, we started with a very broad "law" – Newton's second law – and then translated this in the case of assuming only (constant) gravitational attraction. This model can be completely solved for predictions. The next tasks depend on many things including the goals of the model, how the predictions compare to observations; however, for all good models, there are testable predictions.

In this case, we could predict the maximum height of an object that is thrown, the time at which the object reaches that height, when the object hits the ground, etc. But what if we talk to an experimentalist and we are asked what the terminal velocity is? This is the constant speed that an object reaches when falling. But a look at this model indicates that there is no such thing. In fact, this model assumes the velocity increases like the square of the time.

As modelers, we have to ask what is wrong – our prediction does not match observations. A little exploration uncovers the trouble. There are many other forces that act on the object. There is air-resistance, coriolis force, even forces at the atomic scales. We have to consider which to include and which to exclude. Do we need to include relativistic effects? This is a commonly accepted model of motion that appears at odds with Newton's model.

There is no complete answer for this. In principle all forces need to be accounted for, but this is like a 1-1 scale map. It does not really help matters. But we can work at estimating which forces matter the most. In this case, unless the velocity gets very large, or the mass of the object is tiny or huge, the main forces that matter when considering terminal velocity is drag. We can then develop a model for drag and go back to our experimental colleague and compare our new, revised prediction with their observations. This loop between observations leading to models leading to predictions leading to revised observations is one of the most successful ways for science to progress. Modeling is a crucial step to this leading to quantitive questions (how fast does an object have to move before constant acceleration deviates from observations by more than 5%) that drive experimental design and insight.

In this book, we will also consider a less standard view of models. In many situations, we *know* that there are many interactions. Consider a gene network schematic shown in Figure 1.1. Clearly there is a lot going on. For complicated networks, the models get extremely large – think thousands of equations with tens of thousands of parameters. How can we possibly assess such a model?

We will show how to use sensitivity analysis to reduce the model and make simplifications that greatly aid the connection between models and experiments. This has a different philosophy than used to understand projectile motion. In that situation, we had a simple model that may or may not reflect observations. If the predictions are inaccurate, we look for additional terms. We could think of this as a bottom up approach where you start small and add to the model as need. A top down approach takes a large amount of current knowledge and tries to pinpoint the most relevant parts for specific predictions. Both standpoints are valid and useful and we will move back and forth between them but it is very useful to know the differences between them.

An additional issue that mathematical biologists have to grapple with is the lack of universal laws in biology. There are very few principles that can be viewed in a way that is similar to Newton's laws of motion. Instead, we have to work a bit harder to develop principles that can be translated into theoretical models. Some principles are broad – we use conservation laws in many of the chapters. Some are known from other disciples – the law of mass action has been well developed

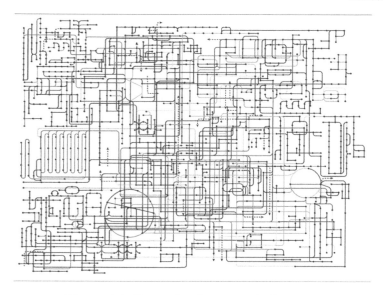

Figure 1.1: Example of a network diagram.

in chemistry. Other principles that guide models is the philosophy that mathematics is portable and that some aspects of models are in some way generic. Biological "switches" often follow the form of Fitzhugh-Nagumo like dynamics. In all of these cases it is very helpful to have a library of models that we know. Therefore many of the examples in the chapters begin with historically important models that set a framework to work within.

For most of the models developed in this book, we will start with simple models and add to them as needed. This has certain specific limitations that we should all be aware of. The main limitation that we will acknowledge is the massive amount of mathematics that can be brought to bear on biological questions. Topics ranging from statistics, calculus, differential equations, machine learning, topology, abstract algebra, graph theory, number theory, geometry, numerical analysis and others have all been used to obtain important insight into biological processes. There is no one book that will cover the diversity of topics. Even so, there is no one researcher who uses all of these methods. For about the past two hundred years, methods from calculus have been extremely effective in translating hypotheses concerning the natural world into mathematical language which can be used to analyze, predict and quantify observations. We will focus almost exclusively on differential equation models – for historical reasons they are the most prevalent but they are also quite flexible.

There are also many, many good books with biological models. These range from undergraduate-level to specialized graduate level texts. It is highly recommended to explore these texts. We list a few here but this is absolutely not a definitive list.

Undergraduate Level

1. The classic book by Edelstein-Keshet, Mathematical Models in Biology [14]. This includes discrete models and continuous ordinary and partial differential equations models.

2. Modeling aimed primarily at biologist's by Day and Otto, A Biologist's Guide to Mathematical Modeling in Ecology and Evolution [53]. A welcoming book focusing primarily on ecology an evolution.

3. A comprehensive book at the calculus level including statistics, basic programming, introductory models and differential equations: Mathematics for the Life Sciences by Bodine, Lenhart, and Gross [6].

4. Essential Mathematical Biology by Britton [9]. This book covers a lot of ground with a variety of models, methods and examples.

Graduate Level

1. Mathematical Physiology by Keener an Sneyd [36] is full of detailed models of an incredible number of biological topics focused on human physiology.

2. J.D. Murray's classic Mathematical Biology. The early, one volume version is quite good but there are more applications and modern models in the two volume editions [49, 50].

3. Lee Segal wrote many modeling books and was one of the original, modern mathematical biologists see [59] and, although not restricted to biology, the classic modeling with CC Lin is chock full of methods and examples [42].

Other authors who have made major contributions to pedagogy (that is teaching, rather than research) include Linda Allen, Avner Friedman, Stacey Smith?, Claudia Neuhauser, Frank Hoppensteadt, Martin Novak, Suzanne Lenhart, and many more.

1.3   Problems

Problems 1.1 *Show that if an object is accelerating due to gravity, there is no maximum velocity.*

Problems 1.2 *Suppose an object is thrown in the air from a height of zero. What is the maximal height that is reached if the initial velocity is 3 ft/s?*

Problems 1.3 *Drag force or air resistance is one of the forces that is neglected in simple projectile motion. One difficulty in modeling air resistance is that direction of the force depends on the velocity. For a projectile fired into the air, the drag force acts in the same direction as gravity when the projectile is moving upward and in the opposite when the object is moving downward.*

(a) *Show that the force $F = -v^2 sgn(v)$ is consistent with this where $sgn(v)$ is 1 if $v > 0$ or -1 if $v < 0$.*

(b) *Consider a ball dropped from rest at height, h. Find the solution to the position as a function of time including air resistance. What is the terminal velocity? Does the concept make sense?*

(c) *How fast does an object need to be traveling before air resistance is 10% of the force of gravity?*

2

Mathematical Background

*This chapter provides an overview of the main tools for analytic stud-
ies. We briefly review differential equations including linear, constant
coefficient equations, separable equations, linearization, and qualitative analysis.*

2.1 Mathematical Preliminaries

Differential equations are relations between unknown function, $y(t)$,
and its derivatives, $F\left(\frac{d^n y}{dt^n}, \frac{d^{(n-1)}y}{dt^{(n-1)}}, ..., \frac{dy}{dt}, y(t), f(t)\right)$. As in algebra, one
of the goals is to determine the unknown that satisfies some constraint
this relationship – for example $F\left(\frac{d^n y}{dt^n}, \frac{d^{(n-1)}y}{dt^{(n-1)}}, ..., \frac{dy}{dt}, y(t), f(t)\right) = 0$. A
few examples are,

$$\frac{dy}{dt} - f(t) = 0,$$

$$\left(\frac{d^2 y}{dt^2}\right)^2 + \frac{d^2 y}{dt^2}\frac{dy}{dt} + \sin(t) = 0,$$

$$\left(\frac{d^3 y}{dt^3}\right)\frac{dy}{dt} + y = e^t.$$

One could naively think of the goal of determining $y(t)$ as some
sort of integration. For the first example above,

$$\frac{dy}{dt} - f(t) = 0,$$

$$\int \frac{dy}{dt}dt = \int f(t)dt,$$

$$y(t) = \int f(t)dt.$$

DOI: 10.1201/9781003316930-2

Sometimes we can undo the derivatives. The solution to the equation $\frac{dy}{dt} - f(t) = 0$ is $y(t) = \int f(t)dt$. Recalling from calculus that until we have specified the domain, these integrals are indefinite and only determined up to a constant. For an equation that has a highest derivative n (referred to as an nth order differential equation), we find n integration constants that need to be determined.

The theory of differential equations is very well developed, especially for the types of models that we will mainly be considering, namely *initial value problems*. For these types of differential equations, the integration constants are determined by providing information at a specific time, Information is typically the value of the function and enough derivatives to determine the constants (remember the initial position and velocity in the projectile motion section). The most widely used alternative to this is to provide information about the unknown function at multiple points. These equations are typically referred to as *boundary value problems* and are more technically challenging.

For initial value problems, broadly speaking as long as all the functions involved are well-behaved (are able to be differentiated a suitable number of times and have no singularities), one can show that there is a unique solution to the equation. We will focus exclusively on these and for the most part initial value problems have unique solutions as long as we prescribe initial values. We can then turn to finding these solutions. To do this, it is often useful to classify initial value problems since certain methods are only useful for certain classes of equations. The reasons to cover analytic methods (that is methods that you can do on paper and completely understand) are two-fold. First, our analysis will lead to insight reality-checks for our models. Second, since we will be doing a lot of of numerical simulations, having certain behaviors in hand will help us interpret our simulations.

Differential equations can be differentiated into two broad categories – *linear* or *nonlinear* depending on whether the relationship $F(\frac{d^{(n)}y}{dt^{(n)}}, \frac{d^{(n-1)}y}{dt^{(n-1)}}, ..., \frac{dy}{dt}, y(t), f(t))$ is linear, with no terms with products of terms containing $y(t)$, or not. Solution techniques and questions about uniqueness of solutions depend a lot on the classification.

2.1.1 Linear

Linear initial value problems can be written as,

$$a_n(t)\frac{d^n y}{dt^n} + a_{n-1}(t)\frac{d^{(n-1)}y}{dt^{(n-1)}} + \dots + a_1(t)\frac{dy}{dt} + a_0(t)y(t) = f(t)$$

$$y(0) = y_0$$

$$\frac{dy}{dt}(0) = y_1$$

$$\vdots$$

$$\frac{d^{n-1}y}{dt^{n-1}}(0) = y_{n-1}$$

If the right-hand-side is zero, this is a homogenous equation and $y(t) = 0$ is a candidate solution, depending on the initial conditions provided.

There are many techniques for analyzing these, but we will focus on one more restrictions. If the coefficients are constant, there is a completely algorithmic way to understand the solution. For a concrete example, we can consider a second order example,

$$a\frac{d^2 y}{dt^2} + b\frac{dy}{dt} + cy = 0 \tag{2.1}$$

$$y(0) = y_0$$

$$\frac{dy}{dt}(0) = y_1.$$

We will come up with a solution by noticing that we already know a function whose derivatives are related to that function. This is the definition of the exponential, $e^{\lambda t}$. So we can guess this as a solution and substitute this into Equation 2.2. This gives us, $a\lambda^2 e^{\lambda t} + b\lambda e^{\lambda t} + ce^{\lambda t} = 0$. Since the exponential is never zero, we can simplify this to find the *characteristic equation*,

$$a\lambda^2 + b\lambda + c = 0.$$

This is an algebraic equation for λ, so we know that there are two values of λ that satisfy the equation,

$$\lambda_\pm = \frac{-b \pm \sqrt{b^2 - 4ac}}{2a}.$$

These are referred to as *eigenvalues*, which is a term that may be familiar if you have had linear algebra. The most important thing to notice at this point is that $y(t) = Ae^{\lambda t}$ is a solution for each of these values of λ and for any value of A. Linear combinations of both of the solutions leads to a solution since the differential equation is linear. In general any solution to the differential Equation 2.2 can be written $y(t) = Ae^{\lambda_+ t} + Be^{\lambda_- t}$. The initial conditions determine A and B.

There are certain things to be aware of. First, what if the values of $\lambda_\pm$ are complex? How do we understand $e^{\lambda_+ t}$ in this case? What happens if $b^2 - 4ac = 0$? For a short version that is applicable to this material, we note that if the eigenvalues are real and distinct, the solutions are exponential and the solutions either exponential increase or decrease to zero. If the eigenvalues are complex conjugate pairs, $\lambda_\pm = R \pm iI$ the solutions can be written as combinations of $e^{Rt}\sin(It)$ and $e^{Rt}\cos(It)$. These solutions oscillate with amplitudes that grow or decay according to whether R is positive, negative or zero.

There is also the case where the eigenvalues are real and repeated. In a differential equations course you learn methods to deal with this. In this book it is less useful to address this. This is what is often referred to as a non-generic case. That is, it requires a precise relationship in the parameters. The point of view that comes from sensitivity and as a practical applied mathematician is that parameters are not absolute, fixed values. They may vary between individual experiments, between individual species, between times of the year or other small differences. As such, it tends to be less important to understand things that require exact values of parameters and we focus on things that occur in wider parameter spaces.

One more useful piece of information is that any higher order differential equation can be written as a system of lower-order differential equations. If we define a new variable, $u = \frac{dy}{dt}$, we can see that $\frac{du}{dt} = \frac{d^2y}{dt^2} = -\frac{b}{a}\frac{dy}{dt} - \frac{c}{a}y$. Equation 2.2 can be written differently,

$$\frac{dy}{dt} = u$$

$$\frac{du}{dt} = -\frac{b}{a}\frac{dy}{dt} - \frac{c}{a}y = -\frac{b}{a}w - \frac{c}{a}y.$$

This can be written very succinctly as

$$\frac{d\mathbf{y}}{dt} = M\mathbf{y}, \tag{2.2}$$

where,

$$M = \begin{bmatrix} 0 & 1 \\ -\frac{c}{a} & -\frac{b}{a} \end{bmatrix}$$

We define $\mathbf{y} = <y(t), u(t)>$.

It turns out that the eigenvalues of this matrix are equivalent to the roots of the characteristic Equation 2.2. We will see that our numerical methods are typically written in this form and, in fact, many of the models come in this form.

That works well for a very restricted class of equations – linear, constant coefficient, homogenous initial value problems. However, most processes in nature are nonlinear. This means that we have to have some way to analyze nonlinear equations. To do this, we will introduce two analytic methods that will be supplemented with direct numerical simulations. The first method illustrates one of the most useful insights in mathematics, namely if we look close enough, most nonlinear processes can be approximated by linear processes. This is used in all areas of mathematics including topology, algebra, differential equations, etc. Any student has seen this in calculus where nonlinear functions are approximated by linear functions using Taylors' theorem. We can do this for differential equations. The second tool is to look at the qualitative behavior of solutions by considering how the derivatives of y control the rate of change of aspects of the graph of y. We will do this in the context of first order equations in this chapter, but in subsequent chapters we will explore the same processes for higher order systems.

2.1.2 Nonlinear Equations

Nonlinear equations are much more difficult to solve analytically. There are several classes that are generally solvable and are focused on in a course in differential equations.

It is typical to write the implicit relationship between $y(t)$ and its derivatives, $F(\frac{d^n y}{dt^n}, \frac{d^{(n-1)}y}{dt^{(n-1)}}, ..., \frac{dy}{dt}, y(t), f(t)) = 0$, in terms of the highest derivative,

$$\frac{d^n y}{dt^n} = G\left(\frac{d^{(n-1)}y}{dt^{(n-1)}}, ..., \frac{dy}{dt}, y(t), t \right).$$

Where G encodes all of the steps needed to isolate $\frac{d^n y}{dt^n}$.

To be very clear, we can start with first-order equations which can be written as,

$$\frac{dy}{dt} = G(y,t). \tag{2.3}$$

These cannot be solved by integrating both sides since the unknown y occurs on both sides. There are some cases where we can *almost* do this though. If $G(y,t) = G_1(y)G_2(t)$ we can rewrite the equation,

$$\frac{dy}{dt} = G(y,t) = G_1(y)G_2(t),$$
$$G_1(y)dy = G_2(t)dt.$$

This is a slight abuse of notation but can be made formally correct. Then integrating both sides with respect to the arguments gives an implicit solution,

$$\int G_1(y)dy = \int G_2(t)dt.$$

Separation of Variables

Consider the differential equation,

$$\frac{dy}{dt} = \frac{t}{y},$$
$$y(0) = 1.$$

We can solve this using separation of variables:

$$\frac{dy}{dt} = \frac{t}{y},$$
$$ydy = tdt,$$
$$\int ydy = \int tdt,$$
$$y^2 = t^2 + c,$$

which gives an implicit solution for y.
The initial condition, $y(0) = 1$, implies that $c = 1$ and we can write an explicit solution $y(t) = \pm\sqrt{t^2 + 1}$.

It should be noted, however, that this does not work in general. Even when it does work, the solution is often only written implicitly and may be quite complicated and unwieldy.

2.2 Linearization

Linearization is one of the major concepts in mathematics. Almost all branches use linearization to approximate the behavior of nonlinear processes. Applied mathematics relies on linearization to a wide extent. We will begin with a brief review from calculus since this is often the first place we see the formal idea of linearization.

Start with a function of an independent variable t, say $f(t)$. Taylors' theorem states that under some conditions on f, we can write $f(t)$ as a polynomial, $P(t) = \sum_i \alpha_i t^i$. Moreover, there is an interval about any point in the domain of f (where f obeys certain restrictions), where $f(t) = P(t)$. This is a remarkable result that provides insight into how to integrate and differentiate a range of functions, since we know that integration and differentiation of polynomials has a predictable patterns. But how do we determine $P(t)$? Taylors' theorem states that, near any point, t, we can write $f(t)$ as,

$$
\begin{aligned}
f(t) &= f(a) + \frac{df}{dt}(a)(t-a) + \frac{\frac{d^2 f}{dt^2}(a)}{2!}(t-a)^2 + \\
&\quad \cdots \quad + \frac{\frac{d^n f}{dt^n}(a)}{n!}(t-a)^n) + \text{error}.
\end{aligned}
$$

Therefore the coefficients of the polynomials are related to the derivatives of f. It also means that as long as t is close to t, $(t-a)^i$ is very small. So we could approximate f,

$$
f(t) \approx f(a) + \frac{df}{dt}(a)(t-a),
$$

where the error term can usually be estimated using the mean value theorem. Graphically, this means that near any point, a nice enough function can be thought of as a line.

This idea can be used to approximate nonlinear differential equations by linear equations. To introduce the topic, we will start with the simplest case of a scalar equation of the form,

$$\frac{dy}{dt} = f(y), \tag{2.4}$$
$$y(0) = y_0,$$

although this idea can be generalized to other forms of differential equations. One of the most important restrictions that we are requiring is that the right-hand-side cannot depend explicitly on time. These equations are referred to as autonomous equations and are the main focus of this text. There are other methods that are used for non-autonomous equations and can be found in textbooks on differential equations (for example [7]).

In this case, the goal is to determine the behavior of the solution. In calculus, we need a specific place to linearize near. Linearization is inherently a local argument and cannot in general be used everywhere. In calculus we linearize at a point. In differential equations, we linearize around a known solution. This seems counter-intuitive at first, since the goal was to find a solution in the first place. However, it is often simple to find some special solutions to differential equations. For example, we can look for solutions that do not depend on time. These are steady-state solutions and are constants that satisfy the differential equation. To find steady-state solutions, $\bar{y}$, we have to solve,

$$f(\bar{y}) = 0.$$

There may be no steady-states, in which case we cannot proceed with the linearization and we have to do something else such as direct numerical simulations. Otherwise, we have at least one solution, $\bar{y}$. We then want to find out how the solution behaves near this steady-state. We define the solution we are studying as,

$$y(t) = \bar{y} + \varepsilon Y(t). \tag{2.5}$$

The function $Y(t)$ is close to $\bar{y}$ as long as $\varepsilon Y(t)$ is "small". We will not be perfectly precise here about how small is small enough but the idea is that understanding the dynamics of $Y(t)$ can provide information about $y(t)$.

To derive an equation for $Y(t)$, we put the solution in Equation 2.5 into Equation 2.4,

$$\frac{dy}{dt} = f(y),$$

$$\frac{d\bar{y} + \varepsilon Y(t)}{dt} = f(\bar{y} + \varepsilon Y(t)),$$

$$\varepsilon \frac{dY(t)}{dt} = f(\bar{y} + \varepsilon Y(t)),$$

$$= f(\bar{y}) + \varepsilon f'(\bar{y})Y(t) + \text{error}.$$

We have used the fact that $\bar{y}$ does not depend on time to simplify the left-hand-side and Taylors' theorem to approximate $f(y)$ with a linear approximation near $\bar{y}$. Notice that this provides an equation for $Y(t)$,

$$\frac{dY(t)}{dt} = f'(\bar{y})Y(t). \tag{2.6}$$

Since $\bar{y}$ is constant and this equation is linear, we know exactly how $Y(t)$ behaves since the solution is $Y(t) = ke^{f'(\bar{y})t}$. If $f'(\bar{y}) > 0$, $Y(t)$ increases and $y(t)$ moves away from $\bar{y}$. On the other hand, if $f'(\bar{y}) <=$, $Y(t)$ decreases and the solution moves towards $\bar{y}$. This says something specific about the steady-state solution, $\bar{y}$. If $f'(\bar{y}) > 0$, we refer to the steady-state as *unstable*. If $f'(\bar{y}) < 0$, $\bar{y}$ is stable. If $f'(\bar{y}) = 0$, the linearization fails and we have to use different arguments to understand the behavior.

We can generalize this idea. If we have a system of n nonlinear differential equations,

$$\frac{dy_1}{dt} = f_1(y_1, y_2, ..., y_n),$$

$$\frac{dy_2}{dt} = f_2(y_1, y_2, ..., y_n),$$

$$...\qquad ...$$

$$\frac{dy_n}{dt} = f_n(y_1, y_2, ..., y_n)$$

we define the steady-state as a vector $\mathbf{\bar{y}} = (\bar{y}_1, \bar{y}_2, ..., \bar{y}_n)$. We look near $\mathbf{\bar{y}}$,

$$\mathbf{y} = \mathbf{\bar{y}} + \varepsilon \mathbf{Y}(t),$$

by inserting this into the differential equations to find,

$$\frac{d\mathbf{Y}}{dt} = J\mathbf{Y},\tag{2.7}$$

We use the notation J since the matrix we have obtained is referred to as the Jacobian. The Jacobian is a useful concept, and for us one of the uses is to shorten some of the calculations. The Jacobian,

$$J = \begin{bmatrix} \frac{\partial f_1}{\partial x_1} & \frac{\partial f_1}{\partial x_2} & \cdots & \frac{\partial f_1}{x_n} \\[2ex] \frac{\partial f_2}{\partial x_1} & \frac{\partial f_2}{\partial x_2} & \cdots & \frac{\partial f_2}{x_n} \\[2ex] \vdots & \vdots & \ddots & \vdots \\[2ex] \frac{\partial f_n}{\partial x_1} & \frac{\partial f_n}{\partial x_2} & \cdots & \frac{\partial f_n}{x_n} \end{bmatrix}\tag{2.8}$$

We show the details for planar systems in the appendix.

2.3   Qualitative Analysis

The last thing that we will review here is a method for understanding the broad behavior of the solutions of differential equations. This does not provide useful information for *quantifiable* predictions. However, it is quite useful for modeling as it can give insight into the general behavior of solutions to models. This often helps diagnose issues with the model that prevent reasonable physical interpretation. Just as linearization does not provide all information (but restricted to starting near a known solution), neither does qualitative analysis.

We will see that qualitative analysis is almost always restricted to scalar or planar systems and it is useful to look at each one separately. We start with scalar equations,

$$\frac{dy}{dt} = f(y).$$

We will again restrict our discussion to autonomous equations and note that we are not imposing initial conditions. Qualitative analysis provides information for the behavior for all initial conditions.

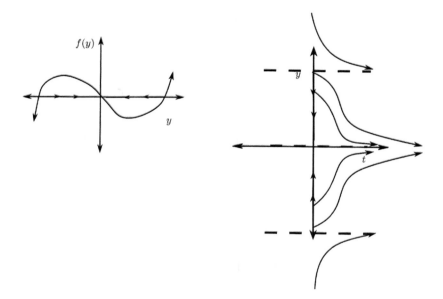

Figure 2.1: Phase line showing how $f(y)$ determines whether y increases or decreases and the qualitative sketch of the solutions, $y(t)$ for different initial conditions.

We can consider the graph of the function $f(y)$ (see Figure 2.1). Where f crosses the y-axis, $f(y) = 0$ which means $\frac{dy}{dt}$ is zero there. That is all roots of $f(y)$ are steady-state solutions. As long as f is nice enough (continuous and differentiable), if the sign of f changes between y_1 and y_2, there is a root. That means that $\frac{dy}{dt}$ has to be of one sign between the roots of f. That means that the solution y is either increasing or decreasing on intervals that do not contain a root. We can sketch the direction that the solution moves on the "phase-line" (see Figure 2.1). This can be used to sketch the solution of the equation as a function of time.

The argument for planar curves is similar but a bit more involved. Consider the system of equations,

$$\frac{dx}{dt} = f(x,y),$$
$$\frac{dy}{dt} = g(x,y).$$

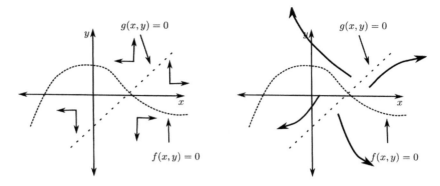

Figure 2.2: Phase plane showing the x and y-nullclines. The direction of increase/decrease in the solution x and y are shown. The same phase plane is shown with a sketch of the trajectories.

The curves $f(x,y)$ and $g(x,y)$ can be drawn in the (x,y)-plane. Above f, x must be increasing. While below f, x is decreasing. Similarly, the evolution of y depends on which side of g we are. We can also see that the solution curve must be tangent to the vector field (f,g). So it is possible to see how trajectories move in the phase-plane – where we think of the solution as a parameterized curve $(x(t), y(t))$ (see Figure 2.2).

2.4 Problems

Problems 2.1 *Classify the following differential equations as linear or nonlinear. Also indicate the order of the differential equation*

(a) $\frac{dy}{dt} + 2y = \sin(t)$

(b) $t\frac{d^2y}{dt^2} + \frac{dy}{dt} = \left(\frac{1}{(1+t^3)}\right) - \left(\frac{3t^2}{(1+t^2)}\right)y$

(c) $-\left(\frac{d^2y}{dt^2}\right)^4 \frac{dy}{dt} = 4$

(d) $4\frac{d^5y}{dt^5} + \cos(t) = 0$

(e) $y\frac{d^3y}{dt^3} - t^2\frac{dy}{dt} + y = 0$

(f) $t^5\frac{d^2y}{dt^2} + t^2\frac{dy}{dt} + y = \sin(t)$

Problems 2.2 *A relation or operator, $F(x)$ is linear if two properties hold: $F(x+y) = F(x) + F(y)$ and $F(cx) = xF(x)$.*

(a) *Show that the derivative operator is linear.*

(b) *Suppose that $y_1(t)$ and $y_2(t)$ both solve the equation,*

$$\frac{d^2y}{dt^2} + \sin(t)\frac{dy}{dt} + \cos(t)y = 10,$$
$$\frac{dy}{dt}(0) = 1,$$
$$y(0) = 1.$$

Show that $y_1(t) + y_2(t)$ is also a solution.

(c) *Show that a linear combination of any two solutions to the differential equation,*

$$\frac{d^2y}{dt^2}\frac{dy}{dt} + y^2 = 0,$$
$$\frac{dy}{dt}(0) = 10,$$
$$y(0) = 10.$$

is not *a solution to the equation.* (Hint: There is no reason to try and solve this equation but see if $y = ay_1 + by_2$ is a solution if y_1 and y_2 are.)

Problems 2.3 *Euler's formula states that $e^{ix} = \cos(x) + i\sin(x)$.*

(a) *Use Euler's identity to show that $e^{R\pm iI} = e^R\cos(I) \pm ie^R\sin(t)$*

(b) *Find two linear combinations of $y_1(t) = e^R\cos(I) + ie^R\sin(t)$ and $y_2(t) = e^R\cos(I) - ie^R\sin(t)$ that involve either $\cos(t)$ or $\sin(t)$.*

Problems 2.4 *The following steps show how to relate the solution to a second order, linear, constant coefficient differential equation with complex eigenvalues to real-valued solutions.*

Consider the differential equation $\frac{d^2y}{dt^2} + 4\frac{dy}{dt} + 5y = 0.$

(a) *Find the eigenvalues.*

(b) *Use Euler's formula to write the two solutions in terms of* $\cos(\alpha t)$ *and* $\sin(\alpha t).$

(c) *Show that the real and complex part of the solutions are themselves solutions.*

Problems 2.5 *The eigenvalues of a* 2×2 *matrix,*

$$A = \begin{bmatrix} a & b \\ b & d \end{bmatrix}$$

Can be found by finding the roots of the determinant of $A - \lambda I,$ *where I is the* 2×2 *identity matrix,*

(a) *Find the eigenvalues of*

$$A = \begin{bmatrix} 2 & 1 \\ -3 & 1 \end{bmatrix}$$

(b) *Show that the eigenvalues of the matrix associated with Equation 2.2 are the same as the roots of the characteristic polynomial of* $\frac{d^2y}{dt^2} = -\frac{b}{a}\frac{dy}{dt} - \frac{c}{a}y.$

Problems 2.6 *Use separation of variables to solve,*

(a)

$$\frac{dy}{dt} = y(y+2),$$
$$y(0) = 1,$$

(b)

$$\frac{dy}{dt} = y(t+3),$$
$$y(0) = 1,$$

Problems 2.7 *(a) Sketch the graph of a nonlinear function $f(x)$, along with the linearization at a point, x_0.*

(b) Use the sketch to show how the linearization provides an approximation of the function value at $x_0 + \delta x$.

Problems 2.8 *Find and classify the steady-states of the following differential equations:*

(a) $\frac{dy}{dt} = y(1 - y)$

(b) $\frac{dy}{dt} = -1 + x^2$

(c) $\frac{dy}{dt} = \sin(x)$.

Problems 2.9 *Suppose we have the planar system of differential equations,*

$$\frac{dx}{dt} = x - xy,$$
$$\frac{dy}{dt} = -y + xy.$$

(a) Find all equilibiria.

(b) Find the Jacobian at the equilibria.

(c) Classify the equilibria.

2.5 Appendix: Planar Example

Many of the examples in this book are planar systems where linearization is very well described. We will briefly cover this and show some of the linear algebra details in this context.

Consider a planar system of equations,

$$\frac{dx}{dt} = f(x,y),$$
$$\frac{dy}{dt} = g(x,y). \tag{2.9}$$

We assume that we have found a steady-state, $(\bar{x}, \bar{y})$ so that $f(\bar{x}, \bar{y}) = g(\bar{x}, \bar{y}) = 0$. We consider the behavior near this solution,

$$
\begin{aligned}
x(t) &= \bar{x} + \varepsilon X(t), \\
y(t) &= \bar{y} + \varepsilon Y(t).
\end{aligned}
$$

Plugging this into the differential equations, using Taylors' theorem for multivariable functions and dropping the nonlinear terms, we find,

$$
\begin{aligned}
\frac{dx}{dt} &= f(x, y), \\
\frac{d\bar{x} + \varepsilon X(t)}{dt} &= f(\bar{x} + \varepsilon X(t), \bar{y} + \varepsilon Y(t)), \\
\varepsilon \frac{dX(t)}{dt} &= f(\bar{x}, \bar{y}) + \varepsilon \frac{\partial f(\bar{x}, \bar{y})}{\partial x} X + \varepsilon \frac{\partial f(\bar{x}, \bar{y})}{\partial y} Y, \\
\frac{dX(t)}{dt} &= \frac{\partial f(\bar{x}, \bar{y})}{\partial x} X + \frac{\partial f(\bar{x}, \bar{y})}{\partial y} Y,
\end{aligned}
$$

for the x-component. Similarly,

$$
\begin{aligned}
\frac{d\bar{y} + \varepsilon Y(t)}{dt} &= g(\bar{x} + \varepsilon X(t), \bar{y} + \varepsilon Y(t)), \\
\varepsilon \frac{dY(t)}{dt} &= g(\bar{x}, \bar{y}) + \varepsilon \frac{\partial g(\bar{x}, \bar{y})}{\partial x} X + \varepsilon \frac{\partial g(\bar{x}, \bar{y})}{\partial y} Y, \\
\frac{dY(t)}{dt} &= \frac{\partial g(\bar{x}, \bar{y})}{\partial x} X + \frac{\partial g(\bar{x}, \bar{y})}{\partial y} Y,
\end{aligned}
$$

for the y-component.

The Jacobian matrix is,

$$
J = \begin{bmatrix} \dfrac{\partial f(\bar{x}, \bar{y})}{\partial x} & \dfrac{\partial g(\bar{x}, \bar{y})}{\partial x} \\ \dfrac{\partial g(\bar{x}, \bar{y})}{\partial x} & \dfrac{\partial g(\bar{x}, \bar{y})}{\partial y} \end{bmatrix}.
$$

For short-hand, we can write this

$$
J = \begin{bmatrix} a & b \\ c & d \end{bmatrix}.
$$

To find the eigenvalues of J we find the roots of the determinant of $J - \lambda I$,

$$\left| \begin{bmatrix} a - \lambda & b \\ c & d - \lambda \end{bmatrix} \right| = 0,$$

$$(a - \lambda)(d - \lambda) - bc = 0,$$

$$\lambda^2 - (a + d)\lambda + ad - bc = 0,$$

$$\lambda^2 - \mathbf{Tr}J\lambda + \mathbf{Det}J = 0.$$

Where $\mathbf{Tr}J$ is the trace of J (the sum of the diagonals) and $\mathbf{Det}J$ is the determinant of J. Therefore, the eigenvalues of the Jacobian are,

$$\lambda +\pm = \frac{\mathbf{Tr}J \pm \sqrt{\mathbf{Tr}J^2 - 4\mathbf{Det}J}}{2}.$$

The components of the Jacobian completely determine whether the real part of the eigenvalues is positive, negative or zero.

3

Introduction to the Numerical Methods

This chapter provides an overview of the numerical platforms that we use throughout the book. We provide some motivation for the two that we have chosen and describe how to start using the programs.

3.1 Introduction

The models that we will use in this book are primarily differential equations. This presents a challenge: How do we understand the models, use them to make useful predictions or provide insight into the biology without several additional years of mathematical training? Typical undergraduate sequences in calculus, linear algebra, differential equations, and scientific computing are useful for understanding the nuances of models. Nonlinearities often put models outside the reach of an undergraduate differential equations class while scientific computing or numerical methods classes often focus on the underlying methods, rather than the applications.

In this chapter, we discuss the two main computer programs/languages that we will use throughout the course. The book will intentionally provide parallel numerical scripts in Python and MATLAB®. Both of these are able to visualize solutions, numerically approximate solutions to a range of equations, and can be used to automate tasks, like sampling, which will be important throughout the book. So why these two packages? Years ago students would have to learn Fortran to be able to solve equations numerically – adding several semesters of course work. Both of these languages are widely used and it is our philosophy that students should have some exposure to all of these languages and then focus on the one that is most comfortable. There are many other languages/platforms that are in use. Languages based on

DOI: 10.1201/9781003316930-3

C-like languages including R are widely used. There are open MPI languages that are able to call both C^{++} and Fortran libraries such as Julia.

This book is relatively agnostic about the language of choice. MATLAB is well developed and has an excellent support system. Python is very flexible and has many workgroups that are designed for helping. They each have some drawbacks as well. MATLAB has a reputation as a "slow" language because it is interpreted. However, there are many ways to speed things up with more experience. Python has a specific culture that can sometimes be overly complicated and difficult to follow since it is so flexible and does not have a centralized structure. R is a relatively complex language to start with and has a steep learning curve. All of these platforms will run into mathematical problems that take an unreasonable amount of time to complete. There should be no examples in this book where this will happen but keep in mind there is no perfect language and many, many people studying how to run more and more complicated problems. The recommendation here is to work with codes in both MATLAB and Python. Eventually, each person will find what is most comfortable.

For those that have never done any programming it might seem strange to use the word "comfortable" – aren't all language uncomfortable? This is definitely true at the beginning. After a bit of work, there are nuances that make certain languages/programs easier for different people. For example, MATLAB m-files do not read indentation. So you can type commands wherever you want. Python is indent-aware and will throw an error (i.e. break and not run) if the indents do not match. This means that the typical Python files look neater and more organized than a typical working m-file. At the same time, MATLAB is more self-contained and does not need any extra libraries to be read in so an m-file written for 2018 MATLAB will work on any machine with 2018 MATLAB. In Python and R, libraries have to be loaded to "teach" the computers Python what certain commands mean. This can occasionally cause issues when sharing files.

All of these are accessible and we will start with essentially no prior knowledge. This chapter is to provide a basic outline of how to access the platform (MATLAB or Python), how to write a program, how to edit a program, and how to run a program. We will consider a few best practices in writing numerical codes in general and in each platform specifically.

One important note about the codes provided in this book. For many chapters we have opted for simpler, less efficient, "home-built" code rather than implementing large scale packages. There are two reasons for this. First, it is much easier to get used to writing and editing scripts with simpler examples – not to mention developing a better understanding of the actual methods used. Second, with all packages there is an issue of upkeep and verification. One cannot assume that the provided code does not have bugs or errors considering companies such as Apple, Google, and Microsoft constantly develop patches to fix codes that cost millions of person hours to develop. Similarly, codes developed for specific projects are not always updates which means it is possible to lose functionality permanently if you don't know how it was written. Some exceptions here are in the later chapters where the implementation is beyond what is expected by readers. The book as chosen a few, well-known, reasonably well-supported implementations; however these are by no means the only methods available. The reader should feel free to explore decide what is the cutting-edge implementation.

3.2   Best Practices in Coding

Writing codes is very much like writing in any language. You would like the writing to be clear, concise, accurate, and readable. One extra consideration is the user interaction with most of these programs. When writing a poem, it is unusual for the author to expect that a reader will add to it or change specific lines or words. Good coding expects this and invites this. So we have to make sure that the files are organized without extraneous parts or things that don't work. In the codes developed in this book, we have made a conscious choice to include some commands that are not the most efficient to ensure they are clear.

3.2.1   Folder Structure

If you open your computer and the desktop is full of files, images, and assorted notes, I would encourage you to start learning file structures. In this book, there will be many different files used. Some will be used for many applications, some will only be used for a single model. Your life will be immeasurably easier if you learn to sort these so you can find them easily and not accidentally delete or overwrite files you are working on. The simplest structure that has some redundancy, but is ordered in a sensible manner might be a folder for the specific course. Within that folder, you have folders for each chapter. Within each chapter folder you have folders for MATLAB and Python and an assignment folder. Within each of the numerical folders you have folders for tasks in the assignment (for example Question 1). There is a high likelihood of redundancy with this since Question 1 might use the same script as Question 2. However, there is much less chance that you accidentally over-write your code answering Question 1 when you start working on Question 2. Typically, for long-lived projects, there are specific times when the file structure is revised. But this should be done as a completely separate task than building the files. It is our expectation that most students using this book will not need to revise their file structure.

One disadvantage to this is that when running scripts, there is the need to change folders. Fortunately in all of our examples, the play button will query this and automatically change folders for you. But it is something to be aware of when sharing codes with other people.

3.2.2 Naming Conventions

All names should be explanatory. This includes folder names, file names, variable names, function names, and code-block names. If you have a file in a folder called `temp1`, it might be junk or it might be the final that you never got around to moving to the correct location. Likewise, a file called `stuff.py` requires a person to open it to see what it is. It is very confusing if you open a file that is supposed to plot a function $f(x,t)$ and x is never used and there is some parameter called g that has a value that is never explained. All of these names should be understandable with very little reader orientation. Consider variable names like `gamma` for a parameter γ. File names like `pred_prey_chapter_1` are easy to understand.

3.2.3 Code Structure

All codes consist of blocks that are typically identified by some commenting convention. In MATLAB it might look like,

```
%%%%%%%%%%%%%%%%%%%%%%%%%%%%%%%%%%%%%%
% Solve the Differential Equation   %
% y' = A y                          %
%%%%%%%%%%%%%%%%%%%%%%%%%%%%%%%%%%%%%%
```

All codes need a header and a body. The header explains what the code is used for, when it was written/edited and any information needed to read and run the code. An example header in Python looks like,

```
"""
Created on Wed Jan 13 12:14:28 2021
Data fitting routine for microbiome data.
Gives current best estimate for parameters
and is the set that SA_microbiome_exc uses
@author: cogan
"""
```

Directly after the header in all our codes will clean our current environment. Computers have a tendency to remember things unless they are explicitly told to forget them. We advise clearing the local and global environment and closing all open figure windows. For MATLAB, this can be done using the commands,

```
clear
close all
```

In Python, we use,

```
from IPython import get_ipython
get_ipython().magic('reset -sf')
```

The body of a code often includes blocks that should be indicated. There are lots of different ways to do this, but we typically use blocked comments. The symbols # and % are used to start comments in Python/R and MATLAB, respectively. So a block might start with

```
###############################
# Visualize all variables     #
###############################
```

in Python and R or

```
%%%%%%%%%%%%%%%%%%%%%%%%%%%%%%%%
% Visualize all variables     %
%%%%%%%%%%%%%%%%%%%%%%%%%%%%%%%%
```

in MATLAB.

3.2.4 Comments

All code should have explicit comments that describe what is going on. These can be done in code blocks or directly after specific lines. These comments should be brief, but self-contained and explain what the specific line or section is doing. You definitely should avoid commenting every line, but anything that reminds the reader what is going on will help remind you what is going on when you look at your codes in 3 months!

3.3 Getting the Programs Running

In this section we describe how to get all programs MATLAB and Anaconda. Provide screen shots of what they look like. Define windows etc.

3.3.1 Python

There are many ways to start using Python. You can download the source files directly and use them from a command line – for example from a terminal window on a Mac or in a c drive on a windows computer. There are also many packages that have been developed to combine an editor where you can write the programs along with a graphical interface. We like the Anaconda package as it is easy to keep updated and well documented. The main package can be found at `https://www.anaconda.com`. When you download and install this, you will be able to open the main window that has many, many options for programs (see Figure 3.1). From here we use the Spyder environment to write and run our Python codes (see Figure 3.2).

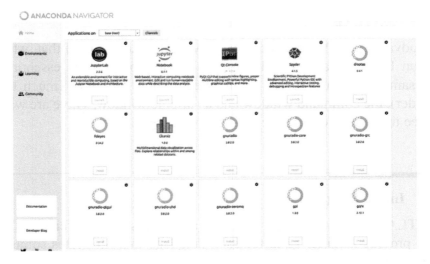

Figure 3.1: Screenshot of the Anaconda Navigator.

Spyder has a lot of buttons that help run and debug programs. These include the green "play" button that runs all the lines scripts and "play/pause" that moves line-by-line. This also gives the path for the programs and other information that helps with navigation etc.

Anaconda also has an environment designed for interactive Jupyter Notebook. This is a nice way to organize blocks of text and codes in one place.

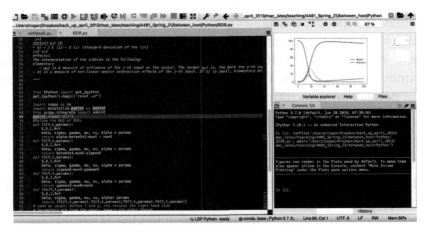

Figure 3.2: Screenshot of the Spyder environment.

3.3.2   MATLAB®

An advantage of Matlab is consistency. Matlab uses a GUI that, while it can be personalized and show different processes, is essentially the same for all users. The main screens show an editor (much like Spyder), the command window and a variable window. There are ways to see the local file structure and other packages.

3.4   Initial Programs

MATLAB and Python treat variables and inputs quite differently. We will provide a starter code for each language that illustrates some of the differences.

In all these platforms we will need to have parameters. These are symbols that we provide specific values for. These are relatively obvious, typing `alpha=1` in all these languages then specifies this value in future commands. Some differences can be seen when running programs. Python does not display anything unless instructed to. MATLAB displays unless instructed not to with a semicolon (;) at the end of a line.

Variables are typically represented as vectors or arrays. In MATLAB, everything is an array. So if `alpha =1`, MATLAB interprets this as a 1×1 array. Just as usual, there are dependent and independent variables. Independent variables require definition while the dependent variables are usually the output of specific commands. To define an independent variable t that lives in the interval $[a, b]$, we will subdivide this interval into n points. The commands to do this for a specific number of points (say $n = 100$) on a specific interval (say $[-\pi, \pi]$)

```
n=100
t=[-pi:2*pi/(n-1): pi]
```

You can check the size of this by typing `size(t)`.

In Python, there is a bit more work. Python is used for a wide variety of applications and often the arrays are not numerical. For example, Python is perfectly happy with an array that is a list of names. To make sure Python has access to the numerical commands we will use, we will add two commands to the header files,

```
import numpy as np
from matplotlib import pyplot
```

To make an array for *t* similar to MATLAB, we can type,

```
t=np.arange(-np.pi,np.pi,1/99)
```

So what is this doing? the `np.*` indicates that * should be interpreted using numpy. If you just type pi, python defaults to thinking of this as letters rather than a fundamental constant. Whereas `np.pi` tells Python that we really mean the constant $\pi = 3.1415....$

If you want to make a dependent variable, it is not terribly different than typical functional notation used in Calculus. In MATLAB you could type

```
f = sin(t),
```

for example. But if you want to multiply variables you have to keep in mind that MATLAB treats everything as an array. So multiplication really means either dot or cross-product. If you want to use the function $f(t) = t^2$, you have to tell MATLAB you don't mean the inner product of the vector *t* with itself. This is done using a special syntax,

```
f = t.^2,
```

where the $.^2$ tells MATLAB you mean to square each element of the array, *t*.

Similarly in Python

```
f=np.sin(t),
```

gives $\sin(t)$.

```
f=t**2,
```

gives $f = t^2$. The $**2$ is Python syntax for "raise to the second power".

Another useful thing is to be able to visualize variables. We will focus on plotting in one dimension, but keep in mind we might want to visualize two or three dimensional functions, scatterplots, barplots, etc. All these programs can do this and we will introduce the syntax as we need it. The basic plotting syntax for MATLAB is `plot(t,f,options)`. We will leave the options for now, but typing `help plot` in MATLAB will give a nice overview of things that can be done. Similarly in Python and R the plotting commands we use are `pyplot.plot(t,f,options)` and `plot(t,f)`, respectively. Note that Python has multiple plotting libraries, but matplotlib is the one that we use and denote it `pyplot`.

The final piece that is important is solving differential equations numerically. This is an incredibly important part of mathematics and there has been tremendous progress in this in the last 60 years. Really, even Newton compiled approximate solutions to differential equations so really, this topic is several hundreds of years old. We are not going to be able to cover everything. Most of the models that we are using are nice enough that we will not run into many issues that the standard differential equation solvers cannot handle. But it is absolutely important to understand that there are many assumptions that go into solving differential equations numerically. It is not difficult to find simple equations that will break all of these methods. This is another reason to have multiple ways to implement numerical methods since the comparison can provide some confidence in the output. But keep in mind, the output from MATLAB or Python can be wrong and sometimes in very serious ways.

So how do we solve differential equations? We need a few things in either MATLAB or Python. Most of the general solvers are built to solve equations in the form,

$$\frac{d\mathbf{y}}{dt} = f(\mathbf{y},t),$$
$$\mathbf{y}(0) = \mathbf{y}_0.$$

So we need a place to store the right-hand-side. We need the initial condition. We also need to tell the programs where to stop. The standard functions in MATLAB and Python will take discrete steps in time and provide estimates for the value of $\mathbf{y}$ at these times. The step-sizes are normally controlled by the internal programs used in the functions called to solve the differential equations. This is to help take large steps if the solution is not changing much and refine the time steps if the function varies a lot. Sometimes this can cause issues – if the solution changes extremely quickly, the step size can get small enough that nothing actually changes. These equations are called *stiff* and special methods need to be used. Also, because the step size may change, it may be difficult to compare solutions quantitatively since the *t*-values may not overlap. This is usually corrected by interpolation – but there are fancier ways.

3.4.1 Differential Equations in Python

There are many, many ways to solve differential equations in Python. We use the package `scipy` that has an integration package and a very flexible solver `solve_ivp`.

A basic script for solving a scalar differential equation only takes a few lines,

```
"""
Created on Wed Dec 22 21:11:32 2021
Basic ODE Script
@author: cogan
"""
from IPython import get_ipython
get_ipython().magic('reset -sf')
import numpy as np
from matplotlib import pyplot
# import plotting library
from scipy.integrate import solve_ivp

def Rhs(t,Y,params):
    r, K = params
    y = Y
    return r*y*(K-y)

IC=.2 # Initial Condition
tstart=0
tstop=4
tspan = np.linspace(tstart, tstop, 100)
params=[.1,100]
y_solution = solve_ivp(lambda t,Y:
Rhs(t,Y,params),
```

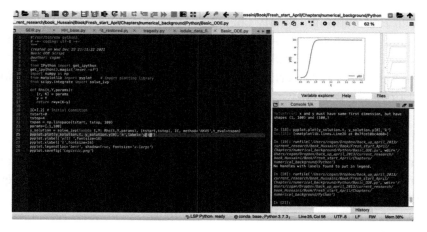

Figure 3.3: Screenshot of the Spyder environment after running the basic differential equation solver.

```
            [tstart,tstop],  IC, method='RK45',
            t_eval=tspan)
pyplot.plot(y_solution.t,
y_solution.y[0],'k',label='y(t)')
pyplot.ylabel('y(t) ',fontsize=16)
pyplot.xlabel('t',fontsize=16)
pyplot.savefig('Logistic.png')
```

A screenshot of the Spyder window shows the result of pushing play is shown in Figure 3.3.

3.4.2 Differential Equations in MATLAB®

Numerical solutions to differential equations in MATLAB have similar requirements as in Python. We have to define the right-hand-sides, the initial conditions and the time interval. MATLAB uses a set of packages (often referred to as the ordinary differential equation (ODE) suite). There are several options including ODE45, ODE23s and others. The most general one is ODE45 and it works for most things that we will need. There are several ways to implement the right-hand-sides, but we will use a method similar in structure to Python.

The only difference is that MATLAB uses two different types of functions. It is typical to include the functions in Python at the top of the script. Often codes in MATLAB are separated into different m-files. One m-file describes the right-hand-side function and the other defines parameters, initial conditions and other inputs, solves the equations and plots them. A sample code for a scalar ODE with the right-hand-side function m-file, in MATLAB is reproduced below. First the file `rhs.m` is defined,

```
function f=rhs(t,Y,params)
%this is the content of the right-hand-side.
%It takes
%inputs t, Y, and params and outputs f
r=params(1);
k=params(2);
y=Y;
f=r*y*(k-y);
```

The differential equation is solve using the m-file, `Logistic_example.m`,

```
%basic Scalar ODE using a m-file call
clear
close all

%define parameters
params=[.2,1];

t_start = 0;
t_stop = 100;
y_0=.1;
[t,y]=ode45(@(t,Y) rhs(t,Y,params), ...
        [t_start t_stop], y_0);

plot(t,y)
xlabel('Time')
ylabel('y(t)')
```

Note that both `Logistic_example.m` and `rhs.m` have to be in the same folder, where MATLAB is currently running.

There are examples in other places where one defines an anonymous function that sits in the `traj_logistic.m` file. This is similar to how Python looks, although the functions for Python are introduced at the top of the script and at the bottom in MATLAB.

3.5   Problems

Problems 3.1 *Complete all the problems in MATLAB.*

(a) Use MATLAB to calculate $5^{(3/2)} + e^{(.1)}$

(b) For loops: Using the syntax:

```
for i=1:.01:10
do something
end
```

Create an m − file that makes a vector $x = [0, .1^2, .2^2, ...]$.

There are at least two ways to build arrays in MATLAB using for loops. One is to "append" to an existing array:

```
a=[]; % creates a memory location for a
for i= 1:.1:1
a=[a, i^2] % appends i^2 to a
end
```

A second way is to enter values into an existing array:

```
a=[]; % creates a memory location for a
for i= 1:10
a(i) = i^2 % puts i^2 into the ith location
end
```

Save the m-file as

```
array_example.m
```

(c) There are two ways of defining functions. You can use an m-file or do it "in-line". I like to use m-file or function calls. The syntax is

```
function f = fun1(x)
f = ***
```

This takes inputs of x and outputs $f(x) = ***$.

Use this to make a function $\frac{\cos(x)}{(1-x)}$.

(d) *Graphing the syntax for a basic plot is* `plot(x,f(x))`,
  where x is a given array and f is a defined function. Use this to plot
  $f(x) = \frac{\cos(x)}{(1-x)}$, where x is the array built in the first problem.

(e) *Differential equations The syntax for solving an ODE,*

$$\frac{dx}{dt} = f(t,x),$$
$$x(0) = x_0$$

requires a few things: a function defining f, an initial condition x_0 and the call for the ODE solver.

An example code starts by defining an m-file that you save as

```
fun1.m
```

containing

```
function f=fun1(t,y)
f=-t*y/sqrt(2-y^2);
```

You also need an initial condition, say $x(0) = x_0 = 1$. Then an interval to solve the equation on, say $t = [0,5]$.

Then in the command window, type

```
[t_out, y_out]=ode45('fun1',[0 5],1);
plot(t_out, y_out)
```

Adjust this to vary the initial condition and interval of solution.

Problems 3.2 *For Python, create a file,* `example.py`. *The header of this should begin with*

```
from IPython import get_ipython
get_ipython().magic('reset -sf')
####################
import numpy as np
from matplotlib import pyplot
# import plotting library
from scipy.integrate import solve_ivp
```

Use this file to edit all the parts of this assignment.

(a) *Use Python to calculate* $5^{(3/2)} + e^{(.1)}$

(b) *For loops: Using the syntax:*

```
for x in my_array :
    do something
```

create a .py-file that makes a vector $x = [0, .1^2, .2^2, ...]$.

(c) *Functions I tend to use inline functions for Python, so in the same Python file, using the syntax,*

```
def f(x):
    f=***
    return f
```

This takes inputs of x and outputs $f(x) = ***$.

Use this to make a function $\frac{\cos(x)}{(1-x)}$.

(d) *Graphing the syntax for a basic plot is*

```
pyplot.plot(t,f)
```

where x is a given array and f is a defined function. Use this to plot $f(x) = \frac{\cos(x)}{(1-x)}$, *where x is the array built in the first problem.*

(e) Differential equations The syntax for solving an ODE,

$$\frac{dx}{dt} = f(t,x),$$
$$x(0) = x_0$$

requires a few things: a function defining f, an initial condition x_0 and the call for the ODE solver.

An example code starts by defining an py-file that you save as

```
fun1.py
```

containing

```
def fun1(t,y):
 f=-t*y/sqrt(2-y^2);
 return f
```

You also need an initial condition, say $x(0) = x_0 = 1$. Then an interval to solve the equation on, say $t = [0,5]$.

```
import numpy as np
from matplotlib import pyplot
 # import plotting library
from scipy.integrate import odeint

# Define a function which
#calculates the derivative\index{derivative}
def Rhs(t, Y,params):
    y=Y
    [a,b]=params
    return a*y-b*t

xs = np.linspace(0,5,100)
params=[-1,.2]
y0 = 1.0  # the initial condition
```

```
ys =solve_ivp(lambda t,Y: Rhs(t,Y,params),
       [tstart,tstop], IC,
       method='RK45',t_eval=tspan)
# Plot the numerical solution
pyplot.plot(xs,ys)
```

Adjust this to vary the initial condition and interval of solution.

3.6   Appendix: Sample Scripts

3.6.1   Python

```
Logistic_example.py:
```

```
"""
Created on Wed Dec 22 21:11:32 2021
Basic ODE Script
@author: cogan
"""
from IPython import get_ipython
get_ipython().magic('reset -sf')
import numpy as np
from matplotlib import pyplot
# import plotting library
from scipy.integrate import solve_ivp
#Define the differential equation
def rhs(t,Y,params):
    r, K = params
    y=Y
    return [r*y*(K-y)]
  # Initial Condition
IC=.2
#Intial time
tstart=0
#Final time
tstop=4
```

```
tspan = np.linspace(tstart, tstop, 100)
#Parameters
params=[.1,100]
y_solution = solve_ivp(lambda t,Y:
rhs(t,Y,params), [tstart,tstop], [IC],
method='RK45',t_eval=tspan)
pyplot.plot(y_solution.t,
                 y_solution.y[0],
                 'k',label='y(t)')
pyplot.ylabel('y(t) ',fontsize=16)
pyplot.xlabel('t',fontsize=16)
#pyplot.savefig('Logistic.png')
```

3.6.2 MATLAB®

```
Logistic_example.m:

%basic Scalar ODE
clear
close all

%define parameters
params=[.1,100];
%Define the time interval
t_start = 0;
t_stop = 4;
%Initial Condition
IC=.2;
%Solving the ODE
[t,y]=ode45(@(t,Y) rhs(t,Y,params), ...
[t_start t_stop], IC);
%plotting the ODE
plot(t,y,LineWidth=2)
xlabel('Time')
ylabel('Y')
%Define the right-hand-side function
function f=rhs(t,Y,params)
r=params(1);
```

```
k=params(2);
y=Y;
f=r*y*(k-y);
end
```

4

Ecology

Ecology is one of the earliest examples of mathematics applied to the life sciences. We describe sequentially complicated models of populations including exponential growth, logistic, Lotka-Volterra, and competitive exclusion. The first sensitivity methods we introduce are scatterplot (e.g. visual) and linear regression.

4.1 Historical Background

Ecology is the study of organisms, their environment, and their interactions. It has been a distinct branch of science since at least the 1850's although studies of populations and their distributions were performed much earlier. In fact, one could argue that much of ancient Greek science revolved around understanding nature and how organisms and their environment fit together. However, the term "ecology" was not coined until 1866 and it was only in common usage by the early 1900's when the British Ecological Society was formed. At that point some scientists referred to themselves as ecologists rather than the previously all encompassing term naturalists and by 1950 the idea of "the study of the structure and function of ecosystems", was firmly established [15, 8].

In some sense, ecology is a specialization of naturalism – the idea that everything stems from a natural order [52, 16]. Early naturalists tried to group organisms in an effort to understand the organization of nature. They developed methods for gathering and classifying observations. There is evidence of the classification during ancient Greece. Aristotle separated plants by size and animals by location (air, water, or land). The question of how to group populations according to similarities and differences, was the fundamental line of inquiry. One of

DOI: 10.1201/9781003316930-4

the difficulties with this is that there was not a systematic way to rank similarities and differences. Large plants are very diverse and it is not clear that this is the best way to divide then up. Even now it is well known that very small variations at the genetic levels can lead to large variations in phenotypes or observable differences. So classifications by apparent similarities may not be the best way to group organisms. It is also difficult to determine the organization of separate types of organisms like mammals versus amphibians. Without specific tests, comparing across types becomes problematic (consider the platypus).

In the 18th century there were major advances in ecology beginning with classification systems proposed by Carlus Linnaeus which is essentially the system still used to a wide extent today. Linnaeus proposed a hierarchical method with kingdoms subdivided into multiple classes, each of which were compose of orders and so on. In some sense this contribution reflects a continuation of the Greek method of classifying with the germ of the scientific revolutionary idea of understanding *why* the classifications are correct and *how* the defining characteristics emerge. Linnaeus proposed to classify by observable details that were not subjective such as the number of male reproductive parts in a flower. By adding complexity and uniformity to the taxonomy, Linnaeus opened the door to identification of mechanisms that connected types of organisms.

In the early 1800's Alexander Humboldt was one of the leading scientists of the day. He travelled extensively and collected prodigious data on species and, more importantly, their habitat. He was the first to describe the range of a species in conjunction with the classification providing immediate insight into the spatial and organism organization. By observing the varying environment with the organisms leads to a host of scientific questions and hypotheses that were not apparent when organisms were organized only by observable features. Simply walking up a mountain and watching how the trees change and abruptly disappear leads to the "why" questions that are fundamental. Humboldt was also keenly interested in the impact of human activity on ecosystems. His observations connected the impact of one species on the development and well-being of another.

The idea of classification naturally turns to questions regarding the cause of distinct classes. Exploring causes naturally leads to the question of interrelationships – who eats whom? Who associates with whom? Who is related to whom? And then, inexorably to theoretical frameworks that can explain why these interrelationships exist. Hypotheses like the food chain that relates predators and prey hierarchically provide a framework to direct observations in a quest to disconfirm or support the hypothesis. As observations accumulate the simplistic linear view of relationships between organisms became untenable. Linear relationships between populations was eventually extended to a web of interconnected chains. In this, species interact directly and also indirectly.

The food web assigns a passive directionality in an ecosystem since energy/nutrient flows from prey to predator. It seemed apparent that it was the prey or herbivores that kept the populations in check. However, this theory was completely altered when Robert Paine performed a classic experiment by creating a food web and removing a single predator species of starfish (*Pisastor ochraceus*) and showing that this drastically altered the community balance. This demonstrated directly that predators help maintain the ecosystem. Further demonstrations lead to the concept of keystone species, one that is more important that others in an ecosystem. This is quite different than the idea of a food chain or even a food web. A food chain would argue that breaking a link and removing one species would affect the species higher up the chain. A web of interactions is more accurate

Humboldt (1769–1859)

He is often referred to as the Father of Ecology. He focused on measurements that pointed toward the interconnectedness of organisms. He measured temperature, rainfall, humidity, and plant life for varying altitudes of Chimborazo in the Andes. Both Lyell's study of geography and Darwin's observational studies were inspired by his published works. The combination of careful observations based on a hypotheses (the interconnectedness of things) led him to be regarded as *the* leading scientist in the world – enough that the New York Times covered the "Humboldt celebration" to mark the 10 year anniversary of his death.

since species connect with multiple species. However, the idea that there were certain organisms that were the key to the ecosystem was transformative.

As ecology has become less of an observational science and more a quantitative science, specific questions can be posed. This is a standard development in a science. To answer the standard How? When? and Why? questions of science, ecology uses a mix of observations (weak inference), experimentation (strong inference) and theory. It is primarily the last of these that is most of interest to us. The primary description in this chapter is the concept of estimating the dynamics of populations in time. This is are pretty narrow version of ecology, but it allows us to look at how models develop and how models can make useful and testable predictions.

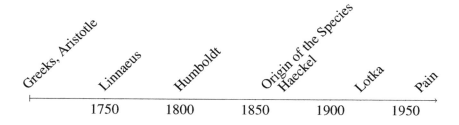

4.2   Single Species Models

The starting point for ecological models is with population models. How do populations grow? What do they respond to? How detailed should these models be? For some species a tremendous amount of detail is known about the processing of nutrient from the environment leading to reproduction. For example, bacterial physiology is relatively well understood and, with careful experimental control, can be essentially completely modeled. For others, details are known at certain scales – typically the large scale group level and the small, cellular level. Complete connection between these scales is very challenging. Regardless, there are common themes that can be explored with population level models.

4.2.1 The Exponential Model

We will start with the simplest model of a growing population. Suppose there are *P*-numbers of people in a population they are reproducing at a rate proportional to the number of people. In words, we might say the population a little in the future is the current population plus the population that is produced, which we assume is proportional to the population. This actually is a pretty complicated assumption. We are clearly assuming something about the actual mode of reproduction, for example. We did not subdivide the population into male and female. We also need to explain a bit about what the rate actually means. If there are only two organisms, it is hard to imagine how they can reproduce at a constant rate. Also, if there is synchronization of the reproduction – for example if there is a breeding season or we are talking about plants with specific flowering/seeding times – we can't really assume a constant rate. Also, the production must depend on how long we wait to observe the population change.

So to be a bit more precise, we are assuming the population is large and diverse with respect to gestation so that on average the change in the population during a small time period. Taking the description from above,

$$P(t + \Delta t) \;=\; P(t) + rP(t)\Delta t, \tag{4.1}$$

where $rP(t)\Delta t$ are the number of offspring produced in the increment of time Δt.

We can do a bit of rearranging and write this as a change in population between t and $t + \Delta t$,

$$P(t + \Delta t) - P(t) \;=\; rP(t)\Delta t. \tag{4.2}$$

Dividing by Δt, yields something that looks a lot like a difference quotient,

$$\frac{P(t + \Delta t) - P(t)}{\Delta t} \;=\; rP(t). \tag{4.3}$$

If we let or observational time get small, we can take the limit $\Delta t \to 0$ and get,

$$\frac{dP}{dt} \;=\; rP(t). \tag{4.4}$$

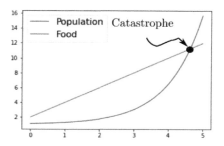

Figure 4.1: Sketch of a Malthusian catastrophe.

This equation might be familiar from calculus as it is often used to *define* the exponential function. That is, the solution to this equation is a function whose derivative is proportional to itself: e^{rt}. Of course, this is not exactly correct because $5e^{rt}$, $-5e^{rt}$, and αe^{rt} all grow at a rate proportional to themselves. We have to specify something to fix this solution. If we assume that we began with exactly P_0 people, we know that $5e^{rt}$ and $-5e^{rt}$ can't be the population. The function αe^{rt} works, but only if $\alpha = P_0$. To see this, just note that if $t = 0$, αe^{rt} becomes $\alpha e^0 = \alpha$. So this model leads to population growth that is exponential.

We could use this to predict the population in the future. These types of observations of population growth within cities began in a formal way in the 18th century. During this period, Europe was becoming more industrial and less rural. As cities became larger, poverty and sanitation became rampant issues of the day. Thomas Malthus attempted to determine natural laws (e.g. theory) to explain the relationship between population growth and poverty. It is clear that exponential growth is not a correct model in general. Populations cannot grow exponentially forever – in fact, ecologists have noted that, while populations vary in time, the overall trend is typically stable over long times [43].

To understand mechanisms that hold exponential growth in check Malthus supposed that population growth and food supply were governed by two different laws, with the population growing exponentially and resources being produced at a constant rate. At some point, the population outruns the food supply and a (Malthusian) catastrophe occurs (Figure 4.1).

4.2.2 The Logistic Model

One criticism of Malthus' theory is the observation that population growth causes resource scarcity. That is, the growth rate of the population should be directly tied to resource availability and similarly resources should be related to the population. In 1838, Pierre-François Verhulst argued that the growth rate should depend on the population level, which implicitly assumes dependence on resource. If the population is low the rate of growth is positive and if it is high the rate is negative. Verhulst proposed the differential equation, which he termed the *logistic* equation,

$$\begin{aligned}
\frac{dP}{dt} &= r(1 - \frac{P}{k})P \\
&= g(P; r, k)P.
\end{aligned} \qquad (4.5)$$

Where we think of the growth rate as $g(P; r, k)$ as a function of the populations rather than a constant as Malthus did.

4.2.3 Analysis

Qualitative analysis for the logistic equation is relatively straightforward. Thinking about the differential equation as defining the derivative of some unknown function, the right-hand-side, $g(P)$ defines regions where the solution $P(t)$ is increasing and decreasing – these are separated by special, constant solutions referred to as equilibria. In fact, the right hand side is a quadratic in P with two roots since $g(P) = 0$ when $P = 0$ or k. Since it is a quadratic (more specifically because it is continuous), we also know that the right hand side cannot change signs except possibly at those two points. So in each of the intervals $(-\infty, 0)$, $(0, k)$, and (k, ∞) the right hand side is either positive of negative. The plot is shown in Figure 4.2.

It is important to note that equilibria solutions are in fact *bona fide* solutions. If $P_0 = 0$ the population remains zero for all time. If $P_0 = k$ then $\frac{dP}{dt} = 0$ and the population remains at the carrying capacity for all time. These are predictions that come from the logistic equation. But what happens if the population is not one of these special equilibria solutions?

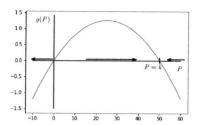 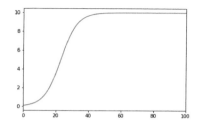

Figure 4.2: Phase line behavior of the trajectories of the logistic equation. It is easy to see that $P = 0$ is unstable while $P = k$ is stable. Any nonzero initial condition will tend toward the carrying capacity, k. State-space trajectory for the logistic equation with $r = .1$ and $k = 10$.

Recall that understanding the behavior near the equilibria or the local answer rests on linearization and can be visualized in the phase-line as was shown in Chapter 2. We linearize using Taylor's theorem around the steady-state $\bar{P} + h(t)$ and find that $h(t)$ satisfies,

$$\frac{dh}{dt} = \begin{cases} rh & \text{if } \bar{P} = 0 \\ -rh & \text{if } \bar{P} = k. \end{cases}$$

Therefore $h(t)$ increases near $P = 0$. If the population starts above $P = k$, it will decrease and below k it will increase. The population always tends to $P = k$. This allows us to determine the long-term behavior of the model is.

Plotting the phase-line along with $g(P)$, we can read off the qualitative behavior (see Figure 4.2).

4.2.4   Predator/Prey – Lotka-Volterra

As ecological hypotheses accumulated and more observations were mode of real ecosystems, the concept of a food web became the dominant paradigm. Here predators eat prey and prey are eaten by predators (and may prey on creatures further down the food chain). A natural next step in modeling populations is to consider one link in the chain instead of treating the interaction between the population and the resource implicitly.

Motivated by this idea Lotka and Volterra developed a two-species interaction model. In fact, Lotka had been interested in biological models that were inspired by chemical models. Differential equation models in chemistry were relatively well grounded at the time and analogies between mass action and population interactions were being explored. At the same time, Volterra was interested in observations of predator fish shortly after WWI. There was a clear variation in the percentage of predator species caught before, during, and after the war. Simultaneously there were variations in the prey species observed. During the war, Adriatic Sea had been fished extensively and presumably this played a role in the variation from a natural, base state. Volterra

---

Working with the computer codes: Scalar

We can do the same calculations using numerical approximations.

$$\frac{dP}{dt} = rP(1 - \tfrac{P}{k}) = f(P)$$

To find the steady-states, we have to solve $f = 0$. In `Steady_state.py` we show an example of how this can be written for functions of two variables. A similar method works for scalar equations. The search intervals matter a lot. Larger intervals lead to longer run-times and the possibility of jumping over steady-states but smaller search intervals restrict the domain. Unfortunately there is not a perfect method for this – one can spend much of calculus learning how to estimate intervals where functions have zeros and we *do not* want to do this here. Instead, we will augment our searching with curve plotting. Simple enough for this example (see Figure 4.2).

We can check whether trajectories near the equilibria agree with our calculations. We do do this by running the code `Traj.py` with an initial population near the steady-states. This also agrees with our notion of stability.

Finally, we can see what happens far from equilibria – run `Traj.py` with any initial value.

posed a two-species model consisting only of predators and prey. This model is equivalent to Lotka's model derived using mass action [2].

The model relates the change in population density of the predator species, H for hunters, with the prey species, P. We assume that the hunters require prey to survive and that otherwise their population decreases. Prey, on the other hand, have an abundance of food, so in the absence of hunters the prey grow at a constant rate. The two species interact with a rate proportional to the product of their densities (this is just mass action). When they meet, the predators convert prey into more predators while the prey are consumed. The equations for this are,

$$\frac{dH}{dt} = \alpha HP - \delta H = f(H,P) \tag{4.6}$$

$$\frac{dP}{dt} = -\beta HP + \gamma P = g(H,P) \tag{4.7}$$

We have to specify initial conditions, $H(0) = H_0$ and $P(0) = P_0$.

4.2.5   Analysis

While this is a fairly simple model it is surprisingly rich in behavior. There are several directions that we could explore at this point. One of the most fruitful is to determine any equilibria and linearize these equations near each one. This provides information about the local (in phase-space) behavior and the long-term behavior. While this is quite useful and often the only "analysis" that can be done, it does not reflect the behavior away from the steady-states – which is often the most useful information for an application. To estimate the behavior in other regions of phase-space, we can explore numerical approximations. Lotka-Volterra equations are sufficiently simple to allow us to extend this to other analytic techniques that do not always generalize.

There are good reasons to try both analytic and numerical methods to understand the model. We will proceed with both of these methods noting only that this motivates using Lotka-Volterra as a canonical example, since we can do this *and* further analysis to help illustrate our methods. We begin with linearization. Remembering the techniques introduced in Chapter 2, we have a reasonable routine to understand the dynamics in certain regions of the variables.

To recap, the process of linear stability requires us to,

1. Find steady-states, $(\bar{H}, \bar{P})$ where $\frac{dH}{dt} = 0$ and $\frac{dP}{dt} = 0$.

2. Linearize the right hand sides: $f(\bar{H} + h(t), \bar{P} + p(t))$ and $g(\bar{H} + h(t), \bar{P} + p(t))$.

3. Examine the linear differential equation for $h(t)$ and $p(t)$.

4. Determine the behavior of h and p. As was introduced in Chapters 2 and 3, we can do this by determining the *eigenvalues of the Jacobian*.

The right-hand-side functions of Lotka-Volterra Equations. $f(H,P) = \alpha HP - \delta H$ and $g = -\beta HP + \gamma P$. There are two equilibria that satisfy,

$$\alpha HP - \delta H = H(\alpha P - \delta) \quad = \quad = 0, \qquad (4.8)$$
$$-\beta HP + \gamma P = P(-\beta H + \gamma) \quad = \quad 0. \qquad (4.9)$$

So that either $H = 0$ or $P = \frac{\delta}{\alpha}$ from Equation 4.9 while $P = 0$ or $H = \frac{\gamma}{\beta}$ from Equation 4.9. The two steady-states are $(\bar{P}, \bar{H}) = (0,0)$, $(\bar{P}, \bar{H}) = (\frac{\delta}{\alpha}, \frac{\gamma}{\beta})$.

Volterra: 1860–1940

Volterra began studying mathematics at a very young age (at age 13, he made contributions to the three body problem). He was a contemporary and well regarded colleague of Pascal, Stokes, and Poincaré. Volterra made seminal contributions in a wide variety of fields. He generalized the theory of functions that paved the way for Hadamard and others to develop theory to address modern physics. The language that he pioneered was formalized and applied by Dirac, for example. He was highly active until the fascist movement in the 1920's. Because of his vocal opposition to Mussolini, he was eventually forced from his university position and fled Italy.

The phase-plane analysis is also quite telling. Recall that for planar systems (two state variables), the behavior is organized by the null-clines ($f = 0$, $g = 0$) and that the equilibria can be seen when these curves intersect. The trajectories (solutions) just follow the direction field. Starting with an initial population of predators and prey, the trajectories in the phase plane form closed curves. In Figure 4.3 we show the behavior in the phase-plane. We also show the parametric curve, (H, P), in the phase-plane corresponds to the one of the curves in state space.

So which version provides the most detail? or the most intuition? It depends greatly on what your goals are. In the context of sensitivity,

---

Working with the computer codes: Steady-states

Now we can run through the same process as before, but with the Lotka-Volterra model. We have differential equations of the form,

$$\frac{dx}{dt} = f_1(x, y)$$
$$\frac{dy}{dt} = f_2(x, y)$$

To find the steady-states , we have to solve two equations simultaneously: $f_1 = 0$ and $f_2 = 0$. In general, this is not simple. So we call the Python code, `Steady_state.py` and we find numerical estimates steady-states. Finding the quantitative values using `Steady_state.py` can be made very accurate, but it takes more time. Try running `Steady_state.py` over larger and larger domains and time the results.

We could get some ideas of where the steady-states might live by just plotting $f_1(x, y)$ and $f_1(x, y)$. But this is a three-dimensional plot (two inputs and an output). It is relatively difficult to visualize this. We can use a simple plotting technique and plot contours of the functions. This is a topographical map of the function. So we can look for zero level sets of f_1 and f_2. Where they intersect are the equilibria.

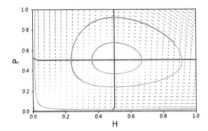 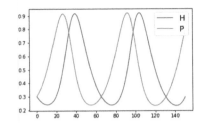

Figure 4.3: Phase-plane and trajectories of solutions. The time-course of a trajectory is also shown. The periodic behavior in the phase-plane corresponds to periodic-in-time solutions.

it might be a bit easier to examine the state-space plot if one wanted to quantify the phase difference in the two curves. If one wanted to know the maximum or minimum either view is sufficient. One thing that is not as apparent in the phase-plane is the period since the trajectories repeat the same curve periodically. But it is important to understand that these are just different views of the solution to the equations, just obtained numerically.

4.2.6 Sensitivity: One at a Time, Scatterplots

Based on the analysis in the previous sections, we can outline a few different questions that can be approached using sensitivity analysis. Obvious targets are things like the steady-states, maximum or minimum values, period, etc. Thinking of this as an ecological model, one might be motivated by determining if the average population is

---

**Working with the computer codes:
Trajectories/Phase-plane**

To understand the behavior more (and to double-check our stability calculations), we use the script `Traj.py` to explore the behavior. If we start "close" to (H_0, P_0) we see behavior consistent with the linear theory – that is $(0,0)$ is unstable and there are rotations around $\left(\frac{\delta}{\alpha}, \frac{\gamma}{\beta}\right)$.

sensitive to variations in parameters. Each of these have strengths and weaknesses – both in interpretation and in calculation. For example, the steady-states can be calculated explicitly, so the sensitivity can be obtained explicitly. However, they have less biological meaning – in fact the population never actually reaches the steady-states since they are not stable, but neutral.

In our case, we will consider the maximum and minimum values within a window of time. Since the period depends on the parameters, and we are varying the parameter, we will let the window be a parameter so that we can determine whether we need to be careful about this. That is, we will let the sensitivity analysis help point the way to the natural period that needs to be considered. We do the simplest thing that we can think of – fix all parameters except for the parameter of interest. We "roll the dice" (e.g. sample from a distribution) values of this parameter. This gives a full parameter set that we use to calculate the solution to the system of equations. We plot success pairs of the parameter value against the quantity of interest (QoI) – here the maximum of the predator species. This scatterplot provides some inference into the how the output (QoI) depends on the input. This is some notion of sensitivity – if the scatterplot appears to be flat, there is little sensitivity. If the trend is increasing or decreasing there is some correlation (positive or negative) and the steepness of the trend is an estimate of sensitivity [19].

We will see later that there are at least two major issues with this type of sensitivity measure. The first is that scatterplots only make sense if one parameter is changed at a time. There is no way to examine cross-correlation or synergistic effects. The second difficulty is in quantifying the sensitivity. Calculating trendlines is reasonably simple if the relationship is almost linear, but if it is not, how should we assign a single number to the relationship? Regardless of this, scatterplots do provide some very nice insight in a computationally inexpensive way. This explains why scatterplots are ubiquitous in data analysis. A large majority of graphs used in scientific publications are in the form of scatterplots – so we should not ignore them.

So, on to the problem at hand. How do we use scatterplots to develop insight into the problem of interacting species? As mentioned before, one trick that we will use first is to explore the effect of the length of time on the QoI. This is important, since we are estimating

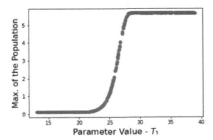

Figure 4.4: Scatterplot of maximum value of the predator population as the end-time of the observation varies. For small times, the QoI is small, then increases up to some the maximum – longer times do not affect this.

aspects of the population – here the maximum value. If we sample for too short a time we might miss the correct estimate. On the other hand, if we sample for a very long time to make sure we can capture the maximum, we are wasting computational resources. We first fix all parameters at some nominal level – this in itself is a complicated task and we *won't* worry about it here, but keep in mind that this does matter. We will solve the equations on the time interval $t \in (0, T_1)$. Once we have selected a nominal value, we vary the end time, T_1.

The results of 500 simulations are shown in Figure 4.4 and tell a simple, useful story. If the time is too short – say less than 22 hours, we vastly underestimated the maximum population value. Beyond some fixed time – about 26 hours, we don't see any changes. This makes sense from the periodicity of the solutions. We have found some measure of the initial transient before the periodic solution is reached.

We will fix the time to be 30 (assuming that there is some variation in the time length as the other parameters vary. In fact, we have good reason to assume this. Two of the parameters are rates and help set the time scale. We can make sure that we have not missed anything once we determine how the QoI depends on the parameters. So, we fix all parameters but one and vary that one, moving through the remaining four parameters.

Looking at the scatterplots shown in Figure 4.5 it is relatively simple to see how each parameter affects the maximum value. Both α and γ are positively correlated with the maximal value of the predator population. These are growth rates, γ being the per capita rate of

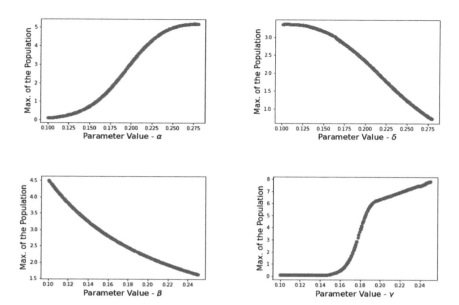

Figure 4.5: Dependence of the maximum value of the predator population for the other parameters. The parameters α and γ are positively correlated with the maximum population while δ and β are negatively correlated.

growth of the predators, so it seems reasonable that increasing this would lead to an increase in the maximum value. Increasing the growth rate of the prey also increases the predator maximum. It is interesting to note that there is an inflection in this correlation. After about 0.18, the rate of increase in maximum value slows down. This would be a place where sensitivity analysis can point toward interesting questions, but does not actually have explanatory power for this type of question. The other two parameters are decay terms. It seems intuitively reasonable for increases in these to lead to decreases in the maximum value of the population.

4.3   Competitive Exclusion

We close this chapter with a small change in the model to include the concept of competitive exclusion . Rather than consider two species interacting in a linear chain, we can think of two species competing for the same resource. Why then, should only one species survive? Niche selection indicates that there are not typically similar species competing for the same resources. This is a fundamental question raised by Darwin, for example. There is evidence that Darwin's consideration of how species survive in niches, coupled with readings of Malthus helped him formulate his version of evolutionary theory.

We select a region of space and insert two populations that have similar foraging strategies. Suppose they consume roughly the same nutrients, have roughly the same motility, reproductive rates, etc. Observing specific species in nature and their niches indicates that these populations will not live happily together. Instead, one will have some advantage and, by exploiting it, will outcompete the other population in a process referred to as exclusion. Beginning in the 1920's controlled experiments began to be performed. In part, this was an effort to reconcile one underlying ecological argument that had been brewing for some time. To what extent does the density of the population affect the behavior? Ecological theories were essentially split between camps that argued that most parameters were density dependent and those that disagreed. To help make a distinction, experiments were performed to see to what extent behaviors depended on density. When the two species were similar and competing for limited resources, it was shown that one species emerged victorious. George Francis Gause [21, 45, 20] formulated this precisely "Two species competing for limited resources can only co-exist if they inhibit the growth of the competing species less than their own growth". It is not a stretch to argue that the principle of competitive exclusion helped organize modern ecology. It provides a broad framework to study how populations interact. If one observes similar species co-existing, the theory of competitive exclusion provides a question: How are these populations evading this interaction.

4.3.1   Model

By combining a Lotka-Volterra type model of population interaction with a Pearl-Verhulst type (logistic) model for population reproduction, a simple model of competition between species with population

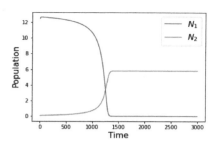

Figure 4.6: An example of exclusion using parameters from Gauses paper [21]. We start with an order of magnitude more of N_1 than N_2, but N_2 exerts more pressure on N_1 than the other way around. In other words, N_2 is more *fit* and eventually N_1 is forced to zero and the competition is won by N_2.

N_1 and N_2 is,

$$\frac{dN_1}{dt} = r_1 N_1 \frac{\kappa_1 - N_1 - \alpha_{12} N_2}{\kappa_1}, \qquad (4.10)$$

$$\frac{dN_2}{dt} = r_2 N_2 \frac{\kappa_2 - N_2 - \alpha_{21} N_1}{\kappa_1}. \qquad (4.11)$$

In his classic paper [21], Gause termed α_{12} and α_{21} the *coefficients of the struggle for existence*. The imagery evoked by Guase is apt. As species struggle for existence, they can do this actively as in predator versus prey or more passively by competing for resources. By considering different species of yeast, Gause studied how the populations grew in isolation and in competition for resources. Later he performed similar experiments on other organisms including paramecium (*Paramecium caudatum* and *P. aureli*).

Using different species of yeast as experimental subjects, Gause studied how the populations grew in isolation and compared this to growth with both species present and in competition for resources. Later he performed similar experiments on other organisms including paramecium (*Paramecium caudatum* and *P. aureli*). Gause also provided estimates of the parameters that reflected his data. Using these parameters and starting with two orders of magnitude more of species N_1, eventually N_2 outcompetes the other population. That is, the "struggle for existence" is won by N_2 (see Figure 4.6).

4.3.2 Analysis

1. Find steady-states , $(\bar{H}, \bar{P})$ where $\frac{dH}{dt} = 0$ and $\frac{dP}{dt} = 0$.

2. Linearize the right hand sides: $f(\bar{H} + h(t), \bar{P} + p(t))$ and $g(\bar{H} + h(t)\bar{P} + p(t))$.

3. Examine the linear differential equation for $h(t)$ and $p(t)$.

4. Determine the behavior of h and p.

Here

$$f_1(N_1, N_2) = r_1 N_1 \frac{\kappa_1 - N_1 - \alpha_{12}N_2}{\kappa_1},$$

and

$$f_2(N_1, N_2) = r_2 N_2 \frac{\kappa_2 - N_2 - \alpha_{21}N_1}{\kappa_2}.$$

The steady-states satisfy,

$$r_1 N_1 \frac{\kappa_1 - N_1 - \alpha_{12}N_2}{\kappa_1} = 0, \qquad (4.12)$$

$$r_2 N_2 \frac{\kappa_2 - N_2 - \alpha_{21}N_1}{\kappa_2} = 0. \qquad (4.13)$$

The axes of the phase plane are nullclines since if either of the populations are zero, they remain zero. So if $N_1 = 0$, the N_1 population cannot change and the same is true for the N_2 population. On these axes, we can understand the behavior. In the absence of one species we have the standard logistic behavior. So the individual populations tend toward their carrying capacity $\kappa_{1,2}$. The other parts of the nullclines come from,

$$\kappa_1 - N_1 - \alpha_{12}N_2 = 0,$$

and

$$\kappa_2 - N_2 - \alpha_{21}N_1 = 0,$$

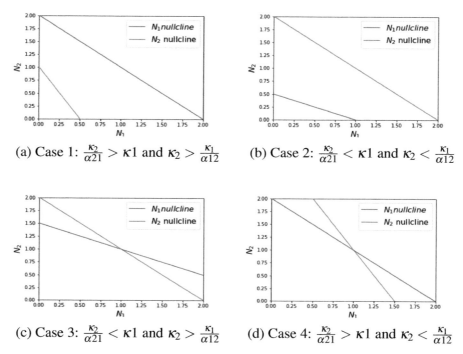

(a) Case 1: $\frac{\kappa_2}{\alpha 21} > \kappa 1$ and $\kappa_2 > \frac{\kappa_1}{\alpha 12}$ (b) Case 2: $\frac{\kappa_2}{\alpha 21} < \kappa 1$ and $\kappa_2 < \frac{\kappa_1}{\alpha 12}$

(c) Case 3: $\frac{\kappa_2}{\alpha 21} < \kappa 1$ and $\kappa_2 > \frac{\kappa_1}{\alpha 12}$ (d) Case 4: $\frac{\kappa_2}{\alpha 21} > \kappa 1$ and $\kappa_2 < \frac{\kappa_1}{\alpha 12}$

Figure 4.7: All cases of competitive exclusion. There are two cases with nonzero steady-states, but only in Case 4 is this stable.

These are lines in the phase-plane. The behavior depends on specifically on the parameters – for example, there are parameter sets for which the steady-states are not in the interior of the first quadrant, that is, one of the species is zero. It is a bit tedious, but not difficult to explore all possibilities. There are four cases, sketched in Figure 4.7. Note that in two of the cases (a and b) there are no other steady-states than either $N_1 = 0$ or $N_2 = 0$, so depending on the initial conditions the trajectories tend toward elimination of one species. When there is a coexistence steady-state, there is one case where the equilibrium is unstable (case c) in which case you again go to a steady-state where one species is eliminated. For the final case, there is a nontrivial steady-state (e.g. a state where neither population is eliminated.

For this case, the equilibrium is stable. But what does this mean biologically? If competitive exclusion is a guiding principle, how can we interpret this observation? One way to get insight it to examine a

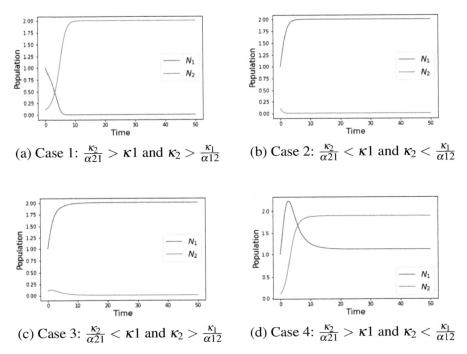

(a) Case 1: $\frac{\kappa_2}{\alpha 21} > \kappa 1$ and $\kappa_2 > \frac{\kappa_1}{\alpha 12}$ (b) Case 2: $\frac{\kappa_2}{\alpha 21} < \kappa 1$ and $\kappa_2 < \frac{\kappa_1}{\alpha 12}$

(c) Case 3: $\frac{\kappa_2}{\alpha 21} < \kappa 1$ and $\kappa_2 > \frac{\kappa_1}{\alpha 12}$ (d) Case 4: $\frac{\kappa_2}{\alpha 21} > \kappa 1$ and $\kappa_2 < \frac{\kappa_1}{\alpha 12}$

Figure 4.8: All cases of competitive exclusion. There are two cases with nonzero steady-states, but only in Case 4 is this stable.

simplified case. What if both populations were exactly the same *except* for their coefficient of struggle? That is, suppose each population had the same carrying capacity and growth rates? Then the condition that has to be satisfied to be in case 4 is that $\alpha_{12} < 1$ and $\alpha_{21} < 1$. This means that the competition terms are small. This begs the question: How small? or Small with respect to what? Now that we have some idea that the magnitude of the competition terms matter, if we go back to the general case (where the populations do not have the same parameters), the inequalities that must be satisfied are $\alpha_{21} < \frac{\kappa_2}{\kappa_1}$ and $\alpha_{12} < \frac{\kappa_1}{\kappa_2}$. So the interaction pressure must be less than the ratio of carrying capacities.

In Figure 4.8 we show results to compare with the nullcline images in Figure 4.7. In Figure 4.9 we show a typical trajectory in the phase-plane. We see that all cases but case 4 lead to exclusion of one species.

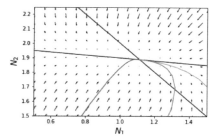

Figure 4.9: Example phase-plane for competitive exclusion for the special case where there is co-existence. The nullclines are shown in red (linear) while specific trajectories are indicated in grey, orange, and blue. Not that all trajectories tend to the steady-state.

4.3.3 Sensitivity: Linear Regression

In this section, we will introduce a sensitivity method related to one-at-a-time scatterplots discussed in Section 4.2.6. It is difficult to rank the parameters in their order of sensitivity by examining the scatterplots directly. This requires some quantification to compare the parameters. Regression is in keeping with the idea of correlation that is visually apparent in the scatterplots but adding a quantitative aspect. Regression means a lot of different things but we will typically talk about it in the informal *curve fitting* approach. The idea is that you have some data and you want to connect them with some function. If you had only two data points you can form a line through the two points. This line could be used to extrapolate to the an independent value outside the data domain (see Figure 4.10 blue star) or interpolate two a value within the domain (see Figure 4.10 green star).

   This begs the question about what to do with more observations? It is impossible to think of their relationship being exactly linear as there so many sources of error and uncertainty. So, what does one do? The easiest quantification follows by *assuming* a linear relationship and asking what is the line that best fits the data. Of course this begs an entirely different question: What constitutes the "best fit"? We will use the relatively robust concept of minimum error, where the error is measured by the sum of the square distances between the model and the data. This is often referred to as the least-squares fitting. This topic

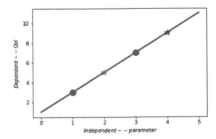

Figure 4.10: Linear extrapolation and interpolation. Given two observations (indicated as red circles), if we assume a linear relationship we can extrapolate (blue star) and interpolate (red star) to get estimates of the QoI away from the observations.

is a deep topic with details that we don't need to understand right now. There are many packages out that will do linear and nonlinear regression . That is they will determine the polynomial (including a linear one) or other nonlinear function that best fits the data. Most sensitivity methods that use regression use linear regression only, since comparing the slopes of regression lines provides robust quantitative comparisons. The major caveat here is that the relationship *must* be linear, or near linear (whatever that means). Remembering the scatterplots Figure 4.7, we know this is not typically true. So we might have to restrict our parameter range, which may weaken our conclusions. There are always trade-offs.

We proceed with quantitative estimates for the competitive exclusion model. But what should the QoI be? Perhaps one of the questions is which parameters are most important for determining which population wins the competition. We define outputs $s_1 = \frac{N_1}{N_1+N_2}$ and $s_2 = \frac{N_2}{N_1+N_2}$. Both s_1 and s_2 are between 0 and 1 (why?). If N_1 is close to zero, s_1 is also. If N_2 is close to zero s_1 will be close to 1. The problem is symmetric, so this is likely to be redundant, but it is simple enough to see what happens. We start as we did with the scatterplots and define intervals for the parameters to be sampled from. We then fix all but one parameter at their nominal value and vary the last one. We put all the selected parameter values into a vector-array, p and the calculated s_1's into a vector-array, S. Running `RegressionLinear.py` finds

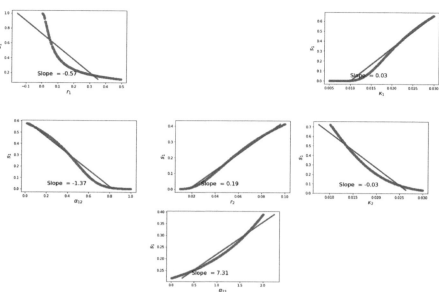

Figure 4.11: Linear regression used to estimate the correlation between the QoI = s_1 for the parameters. Notice that there are several relationships that are *not* linear so it is highly questionable whether this is an appropriate method for estimating the sensitivity.

the best fit line – that is the line with the minimum error. The slope of this line is a measure of the sensitivity. Looking at Figure 4.11, we see very distinct trends. Some of these are easily understandable. For example, as the carrying capacity for population N_1 increases, so does s_1. As the carrying capacity for N_2 increases, s_1 decreases. Similarly the competition terms agree with our intuition. As the pressure on population N_1 due to N_2, α_{12} increases, N_1 decreases. Less clear is the affect of growth rates which appear to be counter intuitive with r_1 negatively correlated with s_1. But even more interesting the scale of the affects. Looking at the estimated slopes, we see that in our parameter regime the competition terms change the output qualitatively more. Since the slopes are several orders of magnitude larger than those for the other parameters, we can conclude that in our parameter regime it is the competition parameters that matter the most – or are the most sensitive.

There is a major caveat here. One can cheat by manipulating the parameter range by choosing a box that is too large and the variance in the output outpaces the variance in the input parameter selections. This can exacerbate the differences in regression lines. One can also unfairly restrict the parameter ranges which can reduce the appearance of sensitivity. The problem is that there is no global method for determining parameter ranges to search over, mainly rules of thumb. It is very useful if these can be restricted by the experimental/biological/-physical system. For example, bacterial growth varies but in planktonic settings this is relatively well controlled and can be estimated. If this is not possible, one should take care to, at the very least, test different parameter sets and this should be clearly identified.

4.4   State of the Art and Caveats

Linear regression is one of the oldest methods to make predictions based on data. One fits a line through the data and uses the equation of the line to predict what will happen beyond where there are observations. The sensitivity methods here use regression in a different way. Rather than using the line to make a prediction, we are using the line to provide a characterization of the QoI/parameter relationship. Sensitivity outputs are almost always scalar and the slope has a nice interpretation in terms of covariation. We could expand our regression to include nonlinear regression; however, this does not lend to single measurements of sensitivity.

   In later chapters, we will expand the one at a time investigation. This winds up being much more important in terms of robustness of our observations. If you visualize the parameter space described by n parameters that each have a minimum and maximum value as an n-dimensional hypercube, one at a time studies are restricted to the surface of the hyper-cube and miss the vast majority of parameter space that is found by varying more than one parameter at a time. This severely constrains our conclusions. In fact, this is probably the single simplest error to avoid in sensitivity analysis. We will immediately address this defect, but it is very important to understand that even if one at a time variations are simple to understand, it is essentially guaranteed that your conclusions are wrong and not robust.

There are several questions other that we have side-stepped that will come back over and over. One of the first that we need to address is our treatment of the parameters. How do we find a nominal parameter set? What does this even mean? It is important to understand that there may be two different answers to this. It is very possible that our conclusions based on our analysis don't actually depend on the exact values of the parameters. For example, in the logistic equation, the value of the carrying capacity, κ doesn't actually play any role in its stability. It is *always* stable no matter what the growth rate is or the carrying capacity as long as they are both positive. These sorts of predictions are often referred to as *qualitative* predictions. Any quantitative prediction requires quantitative understanding of the parameters, but the general question of whether the population tends to the carrying capacity is always true.

Other predictions depend on the parameters. For example, whether species can coexist or not in the competitive exclusion section definitely depends on the parameter values. In the case where our predictions depend on the parameters, we need *quantitative* estimates of the parameters. From a scientific standpoint, these estimates are obtained by comparing models with data. For the most part, we will not address this in this book. It is an important concept but requires much more in depth treatment. Therefore, we will mostly provide parameter estimates that are either found in the literature or used for demonstration purposes.

Another issue that we have not discussed is how the parameters are distributed. We are treating the parameters as if they are random variables and we are "rolling the dice" to obtain sample values. But what if the dice have faces that are not equally represented? If you have a dice with six sides and four of them have one dot, it is far more likely to roll the dice and get a one. The *statistical distribution* of the parameters quantifies the difference in chance of rolling specific values. Distributions that are widely known are Gaussian or normal distribution that is defined by a mean and variance, uniform where all values have equal chance, Poisson, bi-normal, etc. If we know something about the parameters, we might know a good description of the distribution. Typically we may not have this information and it is a choice that the modeler must make. We will usually use uniform or normal – uniform has become the standard if there is no knowledge as it tends to prevent some behavior that is inherent in nonuniform distributions. In later examples, we will consider differences that occur due to this, but it is a current area of research.

4.5 Problems

Problems 4.1 *Use the numerical codes to show that the logistic equation predicts that the population will eventually reach $P = k$ for any positive initial condition.*

Problems 4.2 *Consider the Lotka-Volterra equations,*

$$\frac{dH}{dt} = \alpha HP - \delta H = f(H,P) \tag{4.14}$$

$$\frac{dP}{dt} = -\beta HP + \gamma P = g(H,P) \tag{4.15}$$

(a) *Use one of the numerical codes to show that if the initial population of prey is zero the population of predators goes to zero.*

(b) *Show that starting with only prey leads to unbounded growth.*

(c) *Interpret both of these using the phase-plane. Sketch the trajectories in the (H,P)-plane – Does this agree with the stability of the steady-states?*

Problems 4.3 *The previous problem indicates a difficulty with the simplifying assumptions of the Lotka-Volterra model, namely unbounded growth. Often logistic growth is used when there is evidence that the population is self-limiting. Analyze the logistic Lotka-Volterra model:*

$$\frac{dH}{dt} = \alpha HP - \delta H \tag{4.16}$$

$$\frac{dP}{dt} = -\beta HP + \gamma P(1 - \frac{P}{k}), \tag{4.17}$$

with parameters $\alpha = .2$, $\beta = .1$, $\delta = .2$, $\gamma = .1$, and $k = 2$.

(a) *Find steady-states by solving* $f_1 = 0, f_2 = 0, ... f_n = 0$ *for* $(x_1, x_2, ..., x_n)$.

(b) *Write the solution as a perturbation of the steady-states.*

(c) *Expand and look at the order ε terms.*

(d) *Determine the eigenvalues of the Jacobian.*

(e) *If all eigenvalues are negative, the steady-state is stable.*

Problems 4.4 (a) *Determine the stability of the steady-states of the Lotka-Volterra equations.*

(b) *Demonstrate that the linearization is consistent with the numerical solutions.*

Problems 4.5 *Use scatterplots to determine the sensitivity of the following QoI for the Lotka-Volterra equations:*

1. *The difference in predators and prey population*

2. *Minimum prey*

3. *The period of oscillations. Note that this requires checking if the solutions have converged to a periodic solution and then estimating the period.*

Problems 4.6 *Use linear regression to determine the rankings of the parameters in the competitive exclusion model, when the QoI is the maximum of species N_1. Verify that the behavior is symmetric in species. That is if you compare the rankings with respect to the maximum of N_2 they should be opposite. What is a reasonable interpretation of the sensitivity rankings?*

Problems 4.7 *Demonstrate that the solutions to Lotka-Volterra are periodic – this can be done numerically (see previous problems). Use the period of the output as the QoI and include the initial condition as a parameter. What do you find about the sensitivity of the IC? Why does this make sense?*

Problems 4.8 *(a) Determine the nullclines of the model of competitive exclusion.*

(b) Use these to determine all steady-states. This includes determining when the steady-states exists.

(c) Interpret each steady-state in a biological sense.

5

Within-host Disease Models

This chapter is concerned with disease progression. We consider three different disease types: Bacterial, Viral, and Cancer. We focus on direct sensitivity and how to estimate $\frac{\partial B}{\partial p_i}$.

5.1 Historical Background

Human diseases are incredibly varied in both transmission, progression, and recovery. Some diseases such as sexually transmitted diseases require specific sorts of direct contact that depend on specific behaviors. Other diseases such as Ebola, Zika, and Malaria are spread from active vectors that have their own intrinsic dynamics. Other diagnosis such as the herpes virus are carried within members of the population, but may not present any symptoms unless the host is stressed, meaning some other measure of the state of the host must be included. These may or may not spread during quiescent periods, confounding the models since there is no way to measure the disease-carrying population. Finally, understanding the dynamics of diseases within the host plays an important role in developing treatments since when to treat, how long to treat and how aggressively to treat the disease depends on the hosts response to the disease and treatment.

In this chapter, we will explore different diseases focusing on examples from three categories of diseases: pathological, bacterial, and viral. Each category requires some specific treatment, especially when considering treatment. The pathological category is a mixed group that include cancers that are basically processes that have gone awry, diabetes where specific organs are failing to perform necessary functions, Alzheimers which involves the loss of neuronal functions and the formation of amyloid β plaques. Pathological diseases are typically

DOI: 10.1201/9781003316930-5

treated in very specific methods that have been developed for each individual pathology. Some cancer treatments are often patient focused
employ multiple specific anticancer drugs, chemotherapy, and patient
specific immuno-therapies. For other diseases there is essentially no
treatment.

Bacterial diseases have two-way signaling with the immune system
and can be treated with antibiotics. Bacterial dynamics can depend on
the bacterial species as well as many other factors including whether
the bacteria form a biofilm, whether the bacteria migrate to different
tissues, and whether the host immune system is compromised or not.
The development of drug-resistant strains require new antibiotics that
pose a significant threat to global health. There are currently multiple bacteria that are able to evade all known antibiotics – and are as
untreatable as a bacterial infection was in the 1700's.

Viral diseases tend to evade the immune system better since they
co-opt normal cells to provide a place to reproduce. The advent of
HIV in the 1980's followed by a concerted effort to develop antiviral
treatments and targeted therapy driven by the technology used to drive
the human genome project as led to major advances.

By focusing on the different types of clinical classifications and
understanding, we can begin to understand the role of mathematics in
disease prevention and treatment. For example, even though there are
many antibiotics available, there is widespread concern that bacteria
are evolving to become resistant to treatments. What response should
we have if the goal is to treat diseases and minimize the spread of
resistant strains? How can be treat chronic bacterial infections that are
not caused by genetic changes (e.g. acquiring resistance genes) but are
chronic because of specialized bacterial phenotypes?

More broadly, we would like to use models to help develop preventative measures, treatments and therapies. Can we optimize existing
treatments? If we have multiple drugs available, such as chemothera-
pies and immunotherapies, how should we deploy them? Does timing
matter? Are there methods that minimize side effects? What are the
most important markers to develop patient specific treatments?

Finally, what sorts of theory will help drive our understanding of
the disease cause and progression? Theoretical models can offer a platform to explore a range of hypotheses prior to direct patient contact.

We would like to know what are the most useful outputs from a model. These provide clinicians, drug developers, and facilities information to help point to things that can be changed.

It is interesting that the development of within-host models is far more recent that epidemiological models which we explore in the next chapter. This is due to at least two main reasons. First, it took a long time to begin to understand the causes of diseases. It is hard to imagine a time when the difference between bacterial and viral diseases was not understood, but without this division in infectious diseases, there is almost no way to provide a path to preventing the spread much less developing treatments. By 19th century several competing theories had been developed to understand the nature of diseases. Pasteur developed germ theory by exploring the process of fermentation. He noted that liquids that easily fermented did not behave the same after they were boiled. Since he had already connected the process of fermentation with living organisms, it was a short step to identifying specific bacteria that caused diseases. These were further refined by numerous contemporaries including Koch who studied anthrax. Cancer has been characterized for over a century, but some cancers, such as glioblastomas are very difficult to treat and there has been essentially no improvement in patient outcome in three decades.

The second main reason that models of disease progression is recent is the complexity of the immune system. How does the body actually fight diseases? What are the major components, signals, and processes? Immunology was essentially an experimental subject during the development of vaccines for diseases such as smallpox and cholera. While exploring vaccination, Ehrlich focused on soluble serum or antitoxins (now known as antibodies). Ehrlich followed pioneering work on blood serum by Von Behring and identified chemical mechanisms that formed antibodies against bacterial challenges. Antibodies work by identifying components of invaders (antigens). Antibodies interact with antigens and tag them so that specialized cells can attack and kill them as well as trigger the immune system to train cells to identify and kill the invading cells. Ehrlich and Metchnikoff received the 1908 Nobel Prize in Physiology for their work in identifying the major types and functions of phagocytes.

Different organisms were identified as causal agents of disease and the several mechanisms bodies used to fight infections were identified. However, this did not explain the variation in immune protection. In other words, if antibodies were the bodies method of protecting against disease challenge, what causes or mediates the diversity of antibodies? It was not until the 1950s that the immune system was compartmentalized into innate and active (sometimes referred to as acquired) with immunological memory mediated by specific regulatory machinery including humoral response, active cellular response, and messaging apparatus.

The messaging apparatus has undergone massive revisions from simple chemical activation to lymphocytic interactions and antigen identification, presentation and production dynamics. The most fundamental advance in immunology was the refinement of the concept of clonal adaptation. Specialized cells travel throughout the body, some producing antibodies in response to their environment (B-cells) and some identifying and responding to antigens presented on the surface of cells. The upregulation of production of lymphocytes trained against specific invaders in response to recognition of presented antigens is

---

Ehrlich (1854–1915)

His early career as a chemist coupled with his passion for laboratory research positioned Ehrlich to connect modern cellular theory with the newly developed understanding of molecular biology. This connection introduced the concept of *translational medicine* which aims to connect accurate diagnostic tests that lead directly to treatments. He pioneered the concept that molecular understanding could provide targets for drug development and laid the groundwork for individualized treatment and modern therapeutic development. He developed the first synthetic chemotherapeutic drug and was nominated for (but not awarded) the Nobel prize and was eventually awarded the Nobel prize for his theory of molecular serum. He argued that often stated that in order to be sucessful, one needed the 4 "Gs": Geld ("money"), Geduld ("patience"), Geschick ("skills") and Glück ("luck").

referred to as clonal selection. Basically there is a small population of immune cells that have a library of antigens that they recognize as hallmarks of invaders. If one of the immune cells finds this antigen in the body, it begins to proliferate and lead the charge to attack the invading pathogen.

Until the hypothesis of clonal expansion was solidified in the 1970s there was little identifiable mathematical modeling of within-host dynamics. But this hypothesis leads to quantifiable questions such as how large of an antigen repertoire is required to provide immunity? How do the cells recognize self and non-self antigens? What are the differences between the immune response for viral and bacterial pathogens? How can vaccines be developed and refined? The mathematics relevant to these questions leapt to the forefront during the advent of HIV in the 1980s where the overwhelming nature of the epidemic required concerted effort to understand the process. In 2019, this set the stage for our approach to dealing with COVID-19.

However, this understanding becomes much more complex when cancers are considered. The immune system is geared toward identifying native and non-native cells. It is also responsible for cleaning debris, for example from dead cells after strokes. Cancers have a complex relationship with the immune system. In general, the immune system works to remove cells with mutations that lead to cancers. At the same time, cancers have a strong impact on the immune system and can overwhelm the body's ability to remove mutated cells. Even worse, some breast cancers can be caused by specific macrophages (tumor associated macrophages) that misread the signal from the cancer cells and actually promote growth.

5.2 Pathological: Tumor

For a simple introduction, we will focus on a model proposed for the interaction between a tumor and the immune system [39]. Now, both tumor biology and immune system biology is complicated, so we will simplify this immensely. The drawbacks of this simplification is that we cannot provide detailed insight into any quantitative dynamics. In fact, we likely can't even associate our state variables with biologically measurable processes. The benefit is that the model is quite general and can be adjusted to model many different situations. The basic units of this model represents the disease (in this case a tumor) and the immune response. Recalling that the immune response can be separated into two processes, innate and adaptive (or lots of different terminology). The former is generally always present and designed to respond immediately to invaders in a non-specific way. The adaptive response requires communication where signals recognizing the invaders are used to train the adaptive immune system to respond to a specific threat. The adaptive immune system can be transient or there can be a long term memory that protects the host [63].

5.2.1   Model

This model considers the adaptive immune system. Therefore, we require a few things. First, if there is no tumor the adaptive response should return to some sort of minimal level. This return to a basal state is referred to as homeostasis and ensures that the host does not become over-run by an immune response without a target. We also require the immune response to increase if there is an invader. In the model below, the rate of increase is bounded for all invader levels. For simplicity, we will also expect that the invaders are bounded for all time. We define T as the tumor cells and E as the effector cells – the aggregate of all immune response cells. The word model then looks like,

$$\text{rate of change of E} = \text{production} - \text{decay} + \text{recruitment} - \text{death},$$
$$\text{rate of change of T} = \text{growth} - \text{death}.$$

Tumor cells grow and are killed by effector cells. Effector cells are produced by the body at a constant rate, decay, are recruited by tumor cells (one could lump this into the rate of production by the body, but it is more typical to separate the processes) and die when they remove tumor cells. We just have to describe each of the terms. It is typically a good idea to start simple unless one knows specific observations that need to be accounted for. For example, we have a choice for the tumor growth compartment. We could just assume exponential growth ($\frac{dT}{dt} = kT$). But we have already seen some issues with this and we do know that tumors can't grow forever. We could include nutrient as a limiting factor. That is we could assume $\frac{dT}{dt} = kTf(N)$ for some function of the nutrient. But then we would need an explicit description of the nutrient. This might be useful, but certainly makes the model more complicated. In between these is logistic growth. It leads to bounded behavior but does not require explicit treatment of the nutrient. This is a great place to start, as long as we keep in mind a few things. Logistic growth forces a stable, nontrivial equilibrium. This may actually be inappropriate for some situations. Fortunately, we can always go back and refine the model and this is a place where we know that we are over simplify things.

We will follow previous work done by Kuznetsov and Perelson for the other terms [39],

$$\frac{dE}{dt} = s - dE + pE\frac{T}{g+T} - mET, \tag{5.1}$$

$$\frac{dT}{dt} = rT(1 - \frac{T}{k}) - ET. \tag{5.2}$$

This implies that production of effector cells by the body occurs at a constant rate (probably not a great assumption) and that the reaction between effector and tumor cells reduces each population. This is mass action and the parameters indicate the yield (how many tumor cells are removed by a single effector cell, which has been scaled to one). We will see this in more detail in Chapter 7. Effector decay is at a constant rate.

Recruitment is the most complicated term. First, it takes effector cells to be recruited and they respond to tumor cells. We could have assumed the simplest interaction, pET, but biological observations indicate that there is a maximum rate that effector cells can be recruited so that this rate should be bounded as $T \to \infty$. Again, we could try the model without this and see what happens (see the problems).

We will use the parameter estimates provided as nominal values:

Parameter	Meaning	Value	units
s	Effector production rate	.1181	$\frac{concentration}{time}$
d	Effector decay rate	0.3743	$\frac{1}{time}$
p	Effector recruitment rate	1.131	$\frac{1}{time}$
r	Tumor growth rate	1.636	$\frac{1}{time}$
k	Tumor carrying capacity	0.5×10^3	concentration
g	Effector response scaling	20.19	concentration
m	Effector death from tumor reaction	3.11×10^{-3}	$\frac{1}{concentration}\frac{1}{time}$

5.2.2 Analysis

This is a system of nonlinear equations, which means it is likely to be difficult to solve them analytically. But we can proceed as we did in the previous chapter. Recall the basic steps:

1. Find steady-states by solving $f_1 = 0, f_2 = 0, ... f_n = 0$ for $(x_1, x_2, ..., x_n)$.

2. Linearize the equations.

3. Determine the eigenvalues of the Jacobian.

4. If all real part of the eigenvalues are negative, the steady-state is stable.

To find the steady-states, we need to solve,

$$s - dE + pE\frac{T}{g+T} - mET = 0, \qquad (5.3)$$

$$rT(1 - \frac{T}{k}) - ET = 0. \qquad (5.4)$$

The first equation can be solved for E,

$$E = \frac{s}{d - p\frac{T}{g+T} + mT}$$

and by substituting this into 5.3, we find the ungainly expression,

$$rT\left(1 - \frac{T}{k}\right) - \frac{s}{d - p\frac{T}{g+T} + mT}T = 0. \tag{5.5}$$

Clearly $T = 0$ is one solution. That makes sense since we are not considering how the tumor is initiated. If we do a bit of algebra, we can see how many other solutions exists. We could just plot this, but this comes with a warning that without a decent understanding of how many solutions exists it is easy to miss one so it is better to do a bit of algebra first. We remove the $T = 0$ (corresponding to the steady-state $(E,T) \approx (0.32,0)$) and then eliminate all the fractions,

$$\begin{aligned}
0 &= r(1 - \frac{T}{k}) - \frac{s}{d - p\frac{T}{g+T} + mT}, \\
&= r(1 - \frac{T}{k})(d - p\frac{T}{g+T} + mT) - s, \\
&= r(1 - \frac{T}{k})(d(g+T) - pT + mT(g+T)) - s(g+T) \\
&= S_{steady}(T).
\end{aligned}$$

If you look long enough at Equation 5.6, you can see that it is a cubic in T which means that, in general, there are three nontrivial solutions. In Figure 5.1, we plot $S_{steady}(T)$ to get a visual understanding of the behavior.

For this problem, the Jacobian is relatively simple to calculate analytically:

$$J = \begin{bmatrix} -d + p\frac{T}{g+T} - mT & \frac{pE}{g+T} - pE\frac{T}{(g+T)^2} \\ -T & r - \frac{2rT}{k} - E \end{bmatrix} \tag{5.6}$$

We have determined that there are four equilibria. The tumor free state is unstable and there are two stable equilibria, a low tumor/high effector cell state and a high tumor/low effector state. At this point, there are some options for what to do next. One could note that the model is oversimplified and begin adding complexity, with the aim of incorporating important details to complement the biological investigations. That is, experiments show that compartmentalizing the effector cells this way misses an important feedback loop. Namely, some classes of effector cells (CD 4+) are used to recruit more effector cells.

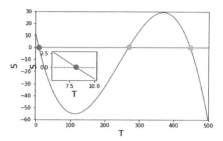

Figure 5.1: Simplified nullclines showing steady-state values of T. There are three intersections. The inset figure shows a zoom-in of the smaller steady-state showing that it is different than $T = 0$.

These "helper" cells have the job of interacting with the tumor cells, which activates them and induces them to produce numerous signaling molecules. One of the most important of these is the specific signal

Working with the computer codes: Steady-states

The equations are again, two dimensional:

$$\frac{dE}{dt} = s - dE + pE\frac{T}{g+T} - mET,$$

$$\frac{dT}{dt} = rT(1 - \frac{T}{k}) - ET.$$

We call the Python `Steady_state.py` and find numerical estimates steady-states. The code `Tumor_effector_ss` has a lot of parts including a numerical Jacobian calculator. We find that there are three equilibria (this is seen in the plot of the three null-clines. We can use the sketch to find starting points for the steady-state solver (either the Steady-State code or it is embedded in the `Tumor_effector_ss` code. Then the numerical estimates of the Jacobian provides an estimate of the stability. For example, we find one of the equilibria has low effector and high tumor burden $((0.75976507, 267.79796146))$ which is stable with the Jacobian having eigenvalues with real parts approximately $(-0.45267824 - 1.69309547)$.

interleukin-2 so some authors include a compartment for the signal that autoinduces recruitment of effector cells.

A different tack could be to use the simplified model to develop and explore treatment methods. One method is to add effector cells (immune therapy). With a fairly simple model such as this one, one can consider adding a bolus of effector cells and try to force the dynamics into the low tumor state (see Homework). The two variable model is not "truth", but is much more tractable. Often, analyzing the simple models provides a road-map for a more complicated model.

5.2.3   Sensitivity: Direct Estimation

We now turn back to sensitivity analysis of our model. In this chapter, we focus on differential methods [11, 64]. One could argue that all sensitivity analysis (SA) methods stem from this method. The question of how much does the output change as the input (parameter) changes is certainly phrased like "rate-of-change" questions in calculus where we define the sensitivity index of a QoI as $S = \frac{\partial QoI(p)}{\partial p}$.

In practice and in historical context, there are plenty of classical examples where analytic methods are used to calculate S, but we will mostly avoid those since it is far more typical for the models to be sufficiently complicated to preclude this type of analysis. In general, nonlinear ODEs do not have closed form solutions, so we have to think of different ways to estimate S. This is the context where the confusion of local versus global might arise. Differential methods are by their nature local since differentiation is a local process (think derivative at a point, or the tangent to a curve at a point). It is important to note that almost all methods of SA are local in this way to some extent – one has to define a region in parameter space. The difference is that the conclusions for differential methods are all evaluated at a point (or a nominal parameter value) and are extended to a neighborhood with caution. In later chapters, we will discuss a different use of the word *global*.

A word of caution here as well. This rapidly becomes computationally difficult for many parameters. What we describe below will be for a single parameter, but can be extended to a vector of parameters.

A useful method for performing local sensitivity using differential thinking to derive a dynamic equation for the sensitivity index of a parameter. Assume that you are given a dynamic model (like Kuznetsov and Perelson) that consists of several ODEs that depend on a parameter p,

$$\frac{dy_i}{dt} = f_i(y_i, t; p), \tag{5.7}$$

with initial conditions,

$$y_i(0) = y_0. \tag{5.8}$$

We define a sensitivity measure $S_i = \frac{\partial y_i}{\partial p}$. This describes how variations in p affect the output y_i. If the exact solution was completely known, this is a useful tool. For example, the logistic equation introduced in Chapter 4, so we can do this explicitly (see the appendix). Typically this is not possible or may be very complicated. Instead, we define new variables, $S_i = \frac{\partial y_i}{\partial p}$ that are the sensitivity measures of each state variable, y_i with respect to a given parameter p. The underlying idea for this method is to consider the dynamics of the index in time, $\frac{dS_i}{dt}$. We can then introduce the dynamics of the state variables (which are known from the ODEs). We take the time derivative of the indices and interchange the time derivative and derivate with respect to the parameter of interest,

$$\begin{aligned} \frac{dS_i}{dt} &= \frac{d}{dt}\left(\frac{\partial y_i}{\partial p}\right), \\ &= \frac{\partial}{\partial p}\left(\frac{dy_i}{dt}\right). \end{aligned} \tag{5.9}$$

Now, interchanging derivatives is not always mathematically valid. It depends on the differentiability of the operations. Here, we will just acknowledge that sometimes this might be a problem, but if the state variables are well behaved and there are not discontinuities is is often correct.

Since $\frac{dy_i}{dt} = f_i(y_i, t; p)$ the chain rule implies,

$$\begin{aligned} \frac{dS_i}{dt} &= \frac{\partial}{\partial p}\left(f_i(y_i, t; p)\right), \\ &= \sum_{j=1}^{n} \frac{\partial f_i}{\partial y_j}\frac{\partial y_j}{\partial p} + \frac{\partial f_i}{\partial p} \end{aligned} \tag{5.10}$$

We just derived a dynamical system (ODE system) that defines how the sensitivity index changes in time. There are two contributions. The first is from how the model depends explicitly on the parameters: $\frac{\partial f_i}{\partial p}$. The other has to do with how the right-hand-sides change implicitly as the dynamics of the state variables are affected by variations in the parameters. This is encapsulated in $\sum_{j=1}^{n} \frac{\partial f_i}{\partial y_j} \frac{\partial y_j}{\partial p}$ Since $\mathbf{S} = (S_1, S_2, ..., S_n) = (\frac{\partial y_1}{\partial p}, \frac{\partial y_2}{\partial p}, ..., \frac{\partial y_n}{\partial p})$, we can write the implicit dependence in terms of the Jacobian,

$$\frac{d\mathbf{S}}{dt} = J\mathbf{S} + \mathbf{f_p},$$

where $\mathbf{f_p} = \left(\frac{\partial f_1}{\partial p}, \frac{\partial f_2}{\partial p}, ..., \frac{\partial f_n}{\partial p} \right)$.

If we have a differential equation for each of the indices, we have to define the initial conditions. We will not show this here, but claim that if no parameters appear in the initial condition of the state equations (Equations 5.8) zero. If the parameter appears explicitly in the initial condition for some state variable y_k, we define $S_i(0) = 0$ unless $i = k$ when $S_k(0) = 1$.

How does this help? Did we just make things harder? It helps because we have tools to estimate the indices. We can just augment our system of ODEs given in Equations 5.7 with those defined in Equations 5.11,

$$\begin{bmatrix} \frac{d\mathbf{y}}{dt} \\ \frac{d\mathbf{S}}{dt} \end{bmatrix} = \begin{bmatrix} \mathbf{f} \\ J\mathbf{S} + \mathbf{f_p}. \end{bmatrix} \tag{5.11}$$

This doubles the size of the system of equations we are solving since $\mathbf{S}$ is the same length as $\mathbf{y}$. We also have to compute the Jacobian matrix at every time-step unless we can precompute this. For systems that are too large this can be computationally difficult (see Problem 5.4). But given those issues, we have a direct method for calculating the sensitivities. This is a good thing! One downside is that it is not as simple to look for QoI's other than the state variables of the system, at least without thinking it through a bit more. There are methods for estimating "feature" sensitivities from direct sensitivities, but these are mainly *ad hoc* (e.g. made up on the spot).

One additional observation that this indicates is that the sensitivities are dynamic in time. In retrospect this is an obvious observation, but it is very apparent in this method. One could ask a question that is more detailed than "What is the most sensitive parameter with respect to the QoI $= y_i$?" Instead, one could ask whether the ranking of the sensitivities change?

An interesting observation can be made for our tumor system. Because there are two state variables (E,T) the state equations are a planar system. This means that we have a lot more theory to rely on in order to understand their dynamics. In particular, we know that for planar systems there are only three ranges of options for the long-term dynamics. The trajectories can tend to a steady-state, they can become unbounded or they can be attracted to a periodic solution. For this system, it can be shown that things tend to one of the two stable steady-states. These can be calculated explicitly (even if it is incredibly messy), but they can be represented $\bar{E}(p_i)$ and $\bar{T}(p_i)$ – as functions of the parameters. So the indices, E_i and T_i for each state variables dependence on p_i could be calculated exactly.

A simple observation in this direction is to start directly on the tumor free steady-state $(s/d, 0)$. Even though this is unstable, if we do not perturb in the T direction, you never leave the steady-state. From the biology this is obvious – the model claims that it requires tumor initiation to upregulate the effectors. Then T is independent of the parameters and should be completely insensitive $T_i = 0$ for all i. One should check that this is correct!

We will use our scripts to study direct sensitivity by augmenting the original dynamical system,

$$\frac{dE}{dt} = s - dE + pE\frac{T}{g+T} - mET = F,$$
$$\frac{dT}{dt} = rT(1 - \frac{T}{k}) - ET = G,$$

with the following system for the index of parameter s,

$$\frac{dS_{E,s}}{dt} = \frac{\partial F}{\partial E}S_s + \frac{\partial F}{\partial T}S_s + \frac{\partial F}{\partial s}$$
$$\frac{dS_{T,s}}{dt} = \frac{\partial G}{\partial E}S_s + \frac{\partial G}{\partial T}S_s + \frac{\partial G}{\partial s}.$$

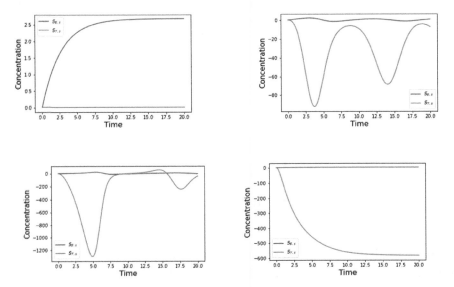

Figure 5.2: Sensitivity indices near each steady-state for the parameter s. The top left and bottom right correspond to no tumor challenge and high tumor challenge, respectively. We see that with no tumor, there is no change in the tumor concentration so T is insensitive to s. For high tumor burden, changes in s tend to suppress the tumor. For the other cases, the approach to the steady-state is periodic, so the impact of s also oscillates.

We can just change our script that we use to solve the differential equations to estimate the solution of the system of four variables. There is a bit of notational difficulty as well – there are indices for each parameter and each QoI (state variable). So we use notation $S_{i,j}$ where i is the QoI (in this case $i = (E, T)$) and j denotes the parameter in question (in this case, s).

In Figure 5.2, we show the sensitivities in time for the outputs E and T with respect to the single parameter s. We show four different results corresponding to the different steady-states where we are initializing the system There are caveats here – this does not rank which parameters are most sensitive, but indicates for which output is s the most sensitive parameter. What is interesting is the comparison between this near different steady-states to explore how the parameter alters the approach to or from the steady state (see Figure 5.3).

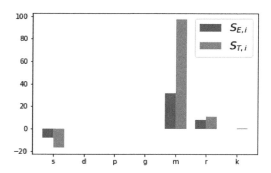

Figure 5.3: Ranking of indices.

We can also examine how the parameter effect the values of T and S at the end of the simulation. This gives some idea of how the parameters suppress or enhance the concentrations of tumor and effector cells. We see that m has the strongest effect and increases both. Looking at the equations, this might not have been intuitive. But we can explain this since m is a measure of how many effector cells are required to eliminate a unit of tumor cells so an increase in m corresponds to lowered effector efficiency which, in turn, leads to higher tumor burden. Increasing the recruitment rate of effector cells (s) leads to lowered tumor burden, which in turn leads to lowered effectors at the end of the simulation time. We could use this to recommend investigating a way to increase the efficiency of the effector cells (lowering m) since that has a large impact on the system.

5.3 Viral: Acute Infection

5.3.1 Model

In this section we consider the case of an acute infection [10]. There is a distinction between this and a chronic disease, even if the models are similar. In particular, acute infections such as influenza, dengue, and Zika have finite duration and are typically resolved within a few weeks whereas chronic infections, such as HIV, persist. The length of time where the disease dynamics plays a role allows for some aspects of the biology such as the production of uninfected cells can be neglected. Additionally, the models are required to exhibit clearance steady-states to be realistic. A generic model is separates the host cells into target cells, T (those that are not infected and possible targets), infected cells, I, and virus, V [10]. When target cells and virus molecules interact the target cells may transition to infected cells. Infected cells can die and produce virus particles which the decay. Using mass action kinetics we can describe the dynamics of each population,

$$\frac{dT}{dt} = -\beta TV, \tag{5.12}$$

$$\frac{dI}{dt} = \beta TV - \delta I, \tag{5.13}$$

$$\frac{dV}{dt} = pI - cV. \tag{5.14}$$

The parameters define the infectivity per virus particle (β), death rate of I (δ), virus production per infected cell (p), and viral decay rate (c). The system has initial conditions T_0, I_0, and V_0. This ignores any latency, where the target cells have to incubate the virus particles. Additionally, this model does not include the immune system. A simple model for the immune response, A, is,

$$\frac{dA}{dt} = kAV + S_A - dA. \tag{5.15}$$

This is a simple model for the immune response to antibodies produced by virus particles since A is produced by the interaction between A and V.

But of course, if we do not include the effects of the immune response on the virus the model has only one-directional effects. That is, the virus signals the immune system without any reciprocal effects. We can couple these by altering 5.14,

$$\frac{dV}{dt} = pI - cV - c_A VA. \tag{5.16}$$

5.3.2 Analysis

We will study the system,

$$\frac{dT}{dt} = -\beta TV, \tag{5.17}$$

$$\frac{dI}{dt} = \beta TV - \delta I, \tag{5.18}$$

$$\frac{dV}{dt} = pI - cV - c_A VA, \tag{5.19}$$

$$\frac{dA}{dt} = kAV + S_A - dA. \tag{5.20}$$

This is a system of nonlinear equations, which means it is likely to be difficult to solve them analytically. However, we can find out a lot about the system by looking at the steady-states and considering the linearization. We need to,

---

Mass Action Kinetics

Mass action kinetics refers to the theory of elementary reactions in chemistry. It is used to translate rate equations (e.g. stoichiometric equations) directly into differential equations. There are many generalizations, but the basic idea can be seen from a general reaction between two molecules, X_1 and X_2, that combine at a rate k^+ to create a product, P. We will also allow the reaction to be reversible with rate k^-,

$$X_1 + X_2 \underset{k^-}{\overset{k^+}{\rightleftharpoons}} P.$$

This turns into differential equations for each of the molecule concentrations,

$$\frac{dX_1}{dt} = -k^+ X_1 X_2 + k^- P,$$

$$\frac{dX_2}{dt} = -k^+ X_1 X_2 + k^- P,$$

$$\frac{dP}{dt} = k^+ X_1 X_2 - k^- P. \tag{5.15}$$

1. Find steady-states by solving $f_1 = 0, f_2 = 0, ... f_n = 0$ for $(x_1, x_2, ..., x_n)$.

2. Determine the linearized system

3. Determine the eigenvalues of the Jacobian

4. Classify the stability of the steady-states.

Often one hears that algebra is "easier than calculus". This does not mean all parts of algebra are equally straightforward. It would be more accurate to say that calculus requires algebra, while algebra does not require calculus so there is some sort of ordering. But it is notoriously difficult to find roots of algebraic equations when the system becomes large – there is no closed form solution like the quadratic formula if the degree of the polynomial is 4 or higher. This does not mean nobody knows what it is, but that it is not possible to find an analytic expression for it, in general.

This is a good lesson in using numerical methods wisely. If you look at the algebraic system,

$$-\beta TV = 0, \tag{5.21}$$
$$\beta TV - \delta I = 0, \tag{5.22}$$
$$pI - cV - c_A VA = 0, \tag{5.23}$$
$$kAV + S_A - dA = 0, \tag{5.24}$$

the first says that steady-states occur only when either $T = 0$ or $V = 0$ – no target cells left of no virus particles. Remember that we are considering acute infections, so the time-scale led us to neglect cell growth. From a purely mathematical standpoint we can reconcile this by arguing in detail about what we mean by long-term. This is a relative description rather than a quantitative description. If we kept going in this direction we would develop the theory of approximations which is often called asymptotics or perturbation theory. While these are very important concepts, they do fall a bit outside the scope of this book. So we will just note that the $T = 0$ case is not relevant for the diseases that we are focusing on. So the steady-state requires $V = 0$. It is simple to see that 5.22 means that $I = 0$. Plugging these into 5.23, we see that it is satisfied and Equation 5.24 gives $A = \frac{S_A}{d}$. That is we expect that

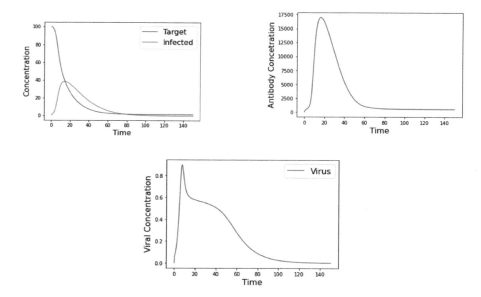

Figure 5.4: Direct numerical solution for the model of an acute infection. We see that the infection peaks around $T = 20$. Of course, we have not specified the time-scale which is set by the rates. In this simulations the parameters are, $\beta = .1181 \; ((m \, hr)^{-1})$, $\delta = .0743 \; (hr^{-1})$, $p = .131 \; (hr^{-1})$, $c = 20.19(hr^{-1})$, $c_A = 3.11 \times 10^{(-3)} \; (hr^{-1})$, $k = 1.636 \; ((m \, hr)^{-1})$, $S_A = .5 * 10^3 \; (m \, hr^{-1})$, $d = 1 \; (hr^{-1})$. Therefore the peak occurs around 20 hours from the onset of the infection.

there is a time-independent solution where there are a given number of target cells, no infected cells, no virus particles and the immune response is at a basal level that is the ratio of production to decay.

For the broader behavior, we can solve the equations numerically for a specific set of parameters. In Figure 5.4, we show the dynamics for each of the variables. Note that the scales are quite different so we have plotted them separately.

5.3.3 Sensitivity Analysis: Feature Sensitivity

Thinking about it for a minute, this is a practically useless result other than noting that once an infection is introduced it will eventually run its course. Instead, it might be useful to ask some more practical questions that a person who is sick might ask. How severe is the sickness

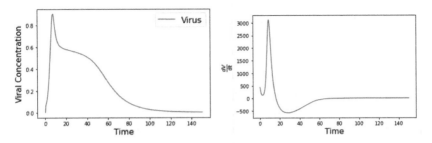

Figure 5.5: The time course of the virus is shown on the left. Post-processing approximation of the rate of change of the viral load in time.

going to be? How long will it last? When can I expect to start feeling better? All of these provide good QoI targets. Examining Figure 5.5, you can see that the basic disease time course from a small infection, is monotonically increasing virus load until a maximum is reached and then monotonic decrease until the virus is cleared. We could just ask which parameters are most important for increasing the rate of change in the viral load. This would tell us what to target during the ramping up of the infection (say to reduce the viral load) as well as during the recovery (to increase the rate of reduction). Mathematically we are interested in $\frac{\partial}{\partial p}\frac{dV}{dt}$. If we optimistically interchange the derivatives we find the sensitivity can be determined from $\frac{d}{dt}\frac{\partial V}{\partial p}$. But this is just the time derivative of S_V, which we have already decided how to calculate. One way to address this, that leans on the computer codes, is to determine an ODE for this so that we can solve it simultaneously. Or we can post-process of solution curves to determine this as well.

For example, we can approximate the derivative of the viral curve shown in Figure 5.4 using a numerical derivative. We have the numerical solution defined at the time points $(t_0, t_1, ..., t_n)$. We can use a difference approximation to estimate the derivative at a particular time,

$$\frac{dV}{dt}\Big|_{t_i} \approx \frac{V(t_{i+1}) - V(t_i)}{t_{i+1} - t_i},$$

as shown in Figure 5.5.

We can use a similar technique to look at the time derivative of the sensitivity. We can use the sensitivity of V with respect to the parameters, using the numerical codes. We then take the numerical derivative of this in time. In Figure 5.6 we show the feature sensitivity of V with respect to S_A.

It is worth noting that numerical differentiation has specific challenges. We are being intentionally naive about this – entire books have been written about methods for differentiating numerical functions or data. Here, we just notice that there are high frequency errors that can be seen in the figures. This is something to be careful about.

Why did we chose to look at this parameter? Looking at the curves, we see that S_A impacts the viral load very early. This parameter is the source of antigens, so this observation actually estimates the most effective time to provide an antigen therapy – that is we know that after the viral load peaks, antigen therapy does not have as much effect on the load. It is interesting to compare the peak in viral load to the peak

---

Working with the computer codes: Steady-state Analysis

Estimates of the steady-states of

$$\frac{dT}{dt} = -\beta TV = f_1, \tag{5.25}$$

$$\frac{dI}{dt} = \beta TV - \delta I = f_2, \tag{5.26}$$

$$\frac{dV}{dt} = pI - cV - c_A VA = f_3, \tag{5.27}$$

$$\frac{dA}{dt} = kAV + S_A - dA = f_4. \tag{5.28}$$

can be found from the Python or MATLAB code `Steady_state`. We already noted that the clearance state is the most relevant. We can ask the stability of this steady-state using the code `Tumor_effector_ss`. In the code `Tumor_Effector_ss.py` we calculate the Jacobian numerically and analytically. We also have a relatively simple numerical Jacobian estimator using a difference approximation in `Jacobian.py`

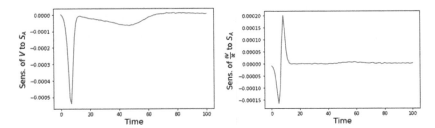

Figure 5.6: An example of using feature sensitivity to examine the impact of variations of S_A on the viral load. The figure on the left is the sensitivity of the viral load on S_A and the figure on the left is the sensitivity of the rate of change of the viral load on the parameter S_A.

in infected cells (Figure 5.4) which happens later. This is a very useful estimate to provide when considering treatments.

5.4 Chronic: Tuberculosis

5.4.1 Model

We now turn to a specific model of a bacterial infection. Rather than describe a model of a "simple" disease where we consider the response of the immune system to an attack, we will discuss a more complicated disease to illustrate how complicated bacterial dynamics can be. Tuberculosis (TB) is caused by the bacterium *Mycobacterium tuberculosis* – mycobacterium refers to the structure of the cell walls. *Mycobacterium tuberculosis* is a gram positive bacterium – which is a classification based on whether the cell walls have peptidologlycan which help the cell walls retain a specific stain. The classification matters, since gram-positive and gram-negative respond very differently to antibiotics. More importantly, tuberculosis has a life-cycle that depends on part of the host immune system [44, 1]. After being introduced into a host through a respiratory pathway, the bacteria reproduce and are identified by the immune system, which primes macrophages to fight the infection. Depending on many different biological processes, some macrophages are unable to digest the bacteria. Instead the bacteria co-opt the macrophage to be a nutrient rich environment where they can reproduce. At some point, the infected macrophage ruptures, releasing bacteria into the host. There are four typical outcomes after the initiation of TB:

1. Primary TB characterized by acute infection with fast or slow progression – noted by increasing numbers of bacteria and immune response.

2. Latent TB characterized by a small, steady number of bacteria and a steady immune response.

3. Clearance and reactivation where the infection is cleared but returns spontaneously.

4. Symptom free latent infection.

To model the disease progression, we can consider a model that separates the bacteria into two classes: active and dormant. The idea is that a portion of the bacteria growing within the body find refuge in macrophages where they remain relatively dormant and evade the immune system where they serve as a source of free bacteria.

The immune system is compressed into two portions. An innate response that is constantly produced and an acquired response that activates due to the free pathogen. The key assumptions are that the dormant bacteria do not reproduce, free bacteria grow exponentially and that there is no lag between the release of dormant bacteria into the free bacterial population. The model state variables are the free bacteria, B; dormant bacteria, Q; and immune response, X.

Going back to the word equation,

Rate of Change of Y = Production – Destruction/decay,

we can write the differential equations for the variables,

$$\frac{dB}{dt} = rB + gQ - (hBX + fB), \tag{5.29}$$

$$\frac{dQ}{dt} = fB - gQ, \tag{5.30}$$

$$\frac{dX}{dt} = a + sX\left(\frac{B}{k+B}\right) - dX. \tag{5.31}$$

We have to provide initial conditions for each of the equations. We start with an initial bacterial infection, $B(0) = B_0$ while we assume there are no dormant bacteria, $Q(0) = 0$. To determine reasonable initial conditions for the immune system, we note that prior to any infection ($B = Q = 0$), the immune system balances constant production at a rate of cells produced per time (a) and decay of cells per time (dX). With $B = 0$, if $X = \frac{a}{d}$, the left hand side of Equation 5.31 is zero and X does not change. This is referred to as a basal state, where the immune system is present, but not excited. We will use this as the initial condition for the immune system.

5.4.2 Analysis

To determine what sorts of steady-states there are, we need to solve the system of algebraic equations

$$rB + gQ - (hBX + fB) = 0, \tag{5.32}$$

$$fB - gQ = 0, \tag{5.33}$$

$$a + sX\left(\frac{B}{k+B}\right) - dX = 0. \tag{5.34}$$

Certainly, Equation 5.33 requires $Q = \frac{f}{g}B$ and we can use this in the other two equations to determine the (B, X) pairs that satisfy,

$$rB - hBX = 0, \tag{5.35}$$

$$a + sX\left(\frac{B}{k+B}\right) - dX = 0.$$

Equation 5.35 has two roots – one when $B = 0$ and the other when $X = \frac{r}{h}$. The former is the disease free state and we know that the steady-state is $(B, Q, X) = (0, 0, \frac{a}{d})$. The latter can be fully determined by replacing $X = \frac{r}{h}$ in Equation 5.36,

$$a + s\frac{r}{h}\left(\frac{B}{k+B}\right) - d\frac{r}{h} = 0.$$

It takes a bit of algebra to determine that $B = \frac{k(dr-ha)}{sr+ha-dr}$ and that there are no other steady-states. We can use the numerical scripts to

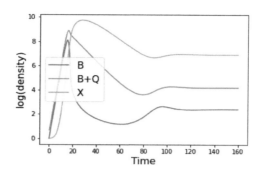

Figure 5.7: Numerical solution of the model of TB. The two bacterial densities are added together to compare the total with the active load.

check the stability of both of these equilibria. If we examine the case where the bacterial population does not vanish (e.g. when the nontrivial steady-state is stable), we can explore the behavior in time see Figure 5.7 and note that one quantity that appears to be feature of the behavior is the peak of the infection.

We might be interested in what the peak is which describes the bacterial load that the body is fighting. We might also be interested in the time at which the peak occurs. This gives some measure of when to expect symptoms to decline. It would also be a useful measurement to compare different treatments. It is simple to see that the time the infection peaks depends on the parameters – this makes some sense since the parameters describe the rate of reproduction, the rate of elimination by the immune system, and the rates of other processes. Just to get a feel for how this works, we took a sample of 100 different parameter values near the nominal values of $f = .5, g = .1, h = 10^3, d = 0.1, s = 1, k = 10^3, a = 0.1, r = .1$). We solve the system numerically and extract the time at which the infection is maximal (see TB.py) and extract the time when the bacterial load is maximum. One way to visualize this is to randomly sample out of a parameter box, evaluate the time at which the maximum occurs and plot a histogram of this data (see Figure 5.8). But what process has the most control of this? If we could identify this, we might be able to propose a target for treatment.

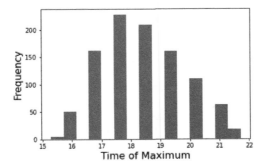

Figure 5.8: Histogram of the time that the active bacteria are maximize. We pulled 1000 random parameter sets, evaluated the differential equations an determined the maximum numerically.

5.4.3 Sensitivity: Relative Change

In this chapter, we are thinking of sensitivity as $\frac{\partial QoI}{\partial p}$. We could always estimate this directly using a discrete approximation. However, if we do this, we are comparing small changes in the parameter to small changes in the QoI, but these might be on very different scales. In fact, this rate of change has dimensions and we need to control for this. The simplest way is to normalize this, so that our measure is,

$$S = \frac{\partial QoI}{\partial p} \times \frac{QoI}{p}.$$

This is often written discretely,

$$S = \frac{\frac{\Delta QoI}{QoI}}{\frac{\Delta p}{p}}.$$

We could define $S = \frac{\frac{\Delta X}{X}}{\frac{\Delta p}{p}}$. This comes from approximating $S = \frac{\partial X}{\partial p}$ which is essentially the definition of local sensitivity.

In Figure 5.9 we show the results for the sensitivity of the maximum time for all parameters. Note that this time is independent of several of the parameters since the sensitivity is zero.

We can see that r, h, and s have strong effects with h positively correlated and the other two negatively correlated.

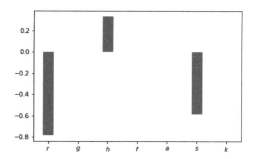

Figure 5.9: Relative sensitivity for the time that the active bacteria are maximal.

5.5 Problems

Problems 5.1 *Consider the within-host model of the tumor effector dynamics:*

$$\frac{dE}{dt} = s - dE + pE\frac{T}{g+T} - mET + \alpha\delta(t_0), \qquad (5.36)$$

$$\frac{dT}{dt} = rT(1 - \frac{T}{k}) - ET. \qquad (5.37)$$

where we have included a source of effector cells at time t_0. Beginning with high tumor, low effector cells $(.1, 500)$, consider how the outcome depends on α and t_0.

(a) For a fixed t_0, systematically vary α and demonstrate that above a certain level, little changes.

(b) Determine whether the time, t_0 plays a role in the affect of α.

(c) Use the differential method to determine the sensitivity of the tumor volume on the two additional parameters α and t_0.

Problems 5.2 *(a) Show that $(\frac{s}{d}, 0)$ is a steady-state of the tumor/effector system: Equations 5.1 and 5.2.*

(b) *Show what happens after adding a small amount of additional effector cells. What does this mean biologically?*

(c) *Show what happens after adding a small amount of tumor cells. What does this mean biologically?*

(d) *Linearize the system near this steady-state and determine the eigenvalues of the Jacobian.*

Problems 5.3 (a) *Look at the differential equation for the stoichiometric equation,*

$$X_1 + X_2 \; \underset{k^-}{\overset{k^+}{\rightleftharpoons}} \; P.$$

What happens if $k^+ \gg k^-$?

(b) *Use the law of mass action to write an equation where the stoichiometric constants are not constant,*

$$2X_1 + X_2 \; \underset{k^-}{\overset{k^+}{\rightleftharpoons}} \; 2P.$$

Hint: This can be written as,

$$X_1 + X_1 + X_2 \; \underset{k^-}{\overset{k^+}{\rightleftharpoons}} \; P + P.$$

(c) *Write the differential equation for the two-step reaction,*

$$X_1 + X_2 \; \underset{k^-}{\overset{k^+}{\rightleftharpoons}} \; P,$$

$$P + C \; \underset{r^-}{\overset{r^+}{\rightleftharpoons}} \; Q,$$

Problems 5.4 *We mentioned that calculating the indices by solving the augmented system might become computationally difficulty. In this problem, we will show that as the size of the system increases, the numerical methods slow down.*

(a) Generate an $n \times n$, random matrix. Use one of the numerical methods to calculate the eigenvalues of this matrix. What happens as n increases?

(b) A Hilbert matrix is an example of an ill-conditioned matrix – which means there is a large difference between the smallest and largest eigenvalue. Repeat the previous part for HIlbert matrices. Both MATLAB and Python have function calls for Hilbert matrices: in Python `scipy.linalg.hilbert(n)` *produces an $n \times n$ Hilbert matrix. In MATLAB* `hilb(n)` *does the same.*

Problems 5.5 *The curves in Figure 5.2 show how the concentration of tumor and effector cells depend on the parameter s and time. Notice that for two parameters, there are not a lot of changes in time while for two other parameters, there is signification variation.*

(a) Examine the trajectories (i.e. solve the equations numerically and plot the variables E and T. Does this help explain why there are oscillations in some of the sensitivity plots? Do the augmented equations explain why some sensitivities are concave up?

(b) Suppose a clinician wanted to know which portion of the system to focus on when developing a treatment. Which parameter(s) would you tell them focus on? Would you make those parameters larger or smaller that the nominal value? Why?

Problems 5.6 *Consider the overly simplified model of within-host dynamics:*

$$\frac{dE}{dt} = s - dE + pET - mET, \tag{5.38}$$

$$\frac{dT}{dt} = aT - nET. \tag{5.39}$$

(a) Find the steady-states.

(b) Use numerical methods to determine if the steady-states are stable or not.

(c) Discuss what biological observations this might support or contradict.

Problems 5.7 *In the appendix, we derive explicit estimates of differential sensitivities for the logistic equation. Compare these estimates with those obtained by scatterplots and augmented SA.*

Problems 5.8 *Use the steady-state and Jacobian codes to explore how each of the steady-states for the TB model changes as key parameters change. Can you use the results from the sensitivity to determine which parameters have the greatest effect on the stability.*

Problems 5.9 *At the end of the chapter, we showed that the direct sensitivity method could be used for certain "feature" sensitivity. In particular, we showed how to estimate which parameters played a dominant role in controlling the rate of clearance of the disease but differentiating the sensitivity measure numerically.*

1. *Adjust the numerical scripts to calculate the feature sensitivity by augmenting a differential equation for the rate of change of the derivative S_V with respect to time. Do you get similar estimates?*

2. *Use either method to estimate the sensitivity of the time to maximum virus concentration and the maximum concentration – discuss how this provides targets to shorten the disease course.*

Problems 5.10 *Consider the differential equations*

$$\frac{dx}{dt} = y,$$
$$\frac{dy}{dt} = -x + y(1 - x^2 - y^2)$$

(a) Show that $(0,0)$ is an unstable steady-state.

(b) Use the numerical codes to show that trajectories that start far enough from $(0,0)$ initially move toward the origin which trajectories that start near the origin move outwards.

(c) Notice that trajectories approach a periodic solution. In the phase plane, show that this orbit is described by $x^2 + y^2 = 1$. This could be done by comparing the trajectories with the unit circle and showing they are very close together.

(d) *On this curve the dynamics can be described by,*

$$\frac{dx}{dt} = y,$$

$$\frac{dy}{dt} = -x$$

Solve this system by differentiating the first equation and using the second equation to find a second order, constant coefficient, linear differential equation that can be solved. What does this show about the restricted trajectories?

Problems 5.11 *Consider the model for tuberculosis in Equations 5.30, 5.31, and 5.31,*

(a) *Show that there is only one steady-state if the bacterial counts are nonzero and determine the stability of the disease free and chronic steady-state.*

(b) *Compare the behavior if the bacteria grow exponentially and logistically by changing r to $r(K - B)$.*

(c) *Change the right-hand-side of 5.31 so that the source term of X depends on the bacteria. How does this change the predictions?*

(d) *Recalculate the sensitivities. Did the rankings change?*

5.6   Appendix

$$\frac{dy}{dt} = ry\left(1 - \frac{y}{k}\right) \tag{5.40}$$

has the exact solution,

$$y = \frac{Ce^{rt}}{\left(1 + \frac{C}{k}e^{rt}\right)}. \tag{5.41}$$

where $y(0) = C$. Interestingly we see that the initial condition can be viewed as a parameter of the system.

Estimating the S_i's is "only" a problem of calculus. It is not terribly difficult to show that

$$
\begin{aligned}
S_C &= \frac{k^2 e^{rt}}{(k + C e^{rt})^2}, \\
S_r &= \frac{C k^2 e^{rt} t}{(k + C e^{rt})^2}, \\
S_k &= \frac{C^2 e^{2rt}}{(k + C e^{rt})^2}.
\end{aligned}
\tag{5.42}
$$

Some things to observe from this are that the sensitivities depend explicitly on time. This is something that we won't discuss here, but sensitivity in time is a useful concept to keep in mind for later. But all sensitivities actually go to zero as $t \to \infty$. This makes sense since we know that for any positive initial condition, the logistic equation tends to a steady-state. If $C < 0$ stranger things can happen! But they are not physical.

But typically it is not possible to find the analytic solution. Instead we turn to approximate techniques. If we do not know the solution the simplest way to estimate S_i is using a difference approximation. $S_i \approx \frac{y_i(p + \Delta p)}{\Delta p}$. This is exact at p and approximates the rate of change/slope of the tangent line, etc. For a single parameter, you can take to realizations of the model (say from a numerical solution) at two different parameter values. Looking at the difference in the output, scaled by the difference in parameters gives an approximation of the indices.

Difference approximation

We can compare the results of the discrete estimate of the sensitivity and the analytic for the logistic equation.
We consider,

$$\frac{dx}{dt} = rx(1 - \tfrac{x}{k}),$$

with the nominal parameter values $r_0 = .1$, $k = 10$ and initial condition, $C_0 = 1$. For simplicity, we will evaluate the sensitivity at some point before the solution has equilibrated, based on our simulations, we use $T = 30$. We find that $S_C \approx 2.2191$, $S_r \approx 66.5716$, and $S_k \approx 0.4457$. That seems to imply that the most sensitive parameter, at that time, is the growth rate – by a lot!
Now we can try the discrete estimate. We have to run `Traj.py` four different times. The first calculates the value of the output at the relevant time, $T = 30$. We will call that Y_30. We then have to vary the parameters, one at a time. To make them comparable, we will vary each parameter by 10% of the nominal value and $S_i \approx \frac{y_i(p+\Delta p)}{\Delta p}$ for the three parameters $p = r, k, C$. We find that the discrete approximation yields $S_C \approx 2.374$, $S_r \approx 64.1062$, $S_k \approx 0.4531$ which is pretty reasonable agreement.

6

Between Host-Disease Models

Between-host models are often referred to as epidemiological models. This is one of the few chapters where we do not specify a particular biological context. The framework of these models is extremely general and include diseases such as the flu, HIV, Covid, and other communicable and vector-born diseases. Sensitivity methods include visual screening methods such as spider plots, tornado plots and cobweb diagrams.

6.1 Historical Background

Mathematics has been used to study disease progression within a population since at least the 1600s when John Graunt used death records from London to develop a method to estimate the risk of dying for different diseases [22]. In fact, it is arguable that alongside ecological models, epidemiological models have been among the earliest models developed as well as those with some of the most lasting affect on peoples daily lives. Insurance costs depends on factors that are, in part, determined by these models. The development of drugs from antibiotics, to antivirals, to vaccines (both in the development of the vaccine and in the use of the vaccine to prevent the spread of diseases) are motivated and assisted using mathematical models. Questions such as where should governments allocate resources for disease prevention such as HIV prevention (needle exchange? education?) and outbreak diseases like Ebola, West Nile virus are influenced by epidemiological models. Mathematicians from Daniel Bernoulli – who introduced a model of smallpox treatments in 1760 – to John Snow – who studied the spread of cholera and used his quantitative theory to locate the source of a cholera outbreak in London to current mathematicians

DOI: 10.1201/9781003316930-6

including those working with governments throughout the world use mathematical models to inform decision-making as well as preventative methods. During the development of modern theory, models have gone from highly disease specific (e.g. focusing on one disease in one region) to relatively general (focusing on theoretical and qualitative results) to back to more complex and specific models that use the general theory as a guide to understanding.

The outbreak and spread of COVID-19 is an example that is occurring at the time of this writing. Predicting the spread of the disease, excess mortality, impact of methods to control the pandemic and even economic cost of the disease all rely on models. Because models rely on specific assumptions each model analysis often produces different predictions. It is the job of modelers to separate out predictions that differ in magnitude, such as the percent of the population that will be infected, from those that do not, such as whether or not a disease will become endemic (i.e. permanently circulating in the population). It is also the job of modelers to assess the most viable targets for controlling the disease. One of the most controversial topics during the Covid pandemic was the extent masks helped prevent the spread of the disease. Part of this controversy was a consequence of evolving understanding of the spread of the disease. Masks help more if the disease is airborne than if it is not. Another source of different quantitative predictions was a lack of knowledge of specific measurements such as how many particles different types of masks can filter. These sorts of questions can be addressed by estimating the parameter sensitivities as well as analyzing aspects of mathematical models.

Another interesting outcome of the global pandemic is the prevalence of terms that are fundamental in epidemiology – terms like "herd immunity" and "R naught (R_0)". It is defined as the number of new infections caused by one infected member of the population over the course of the infection. Intuitively, if R_0 is smaller than one the infection spreads too slowly to maintain or increase in number. This quantity is of fundamental importance in epidemiology since it provides the border between the spread and decline of a disease. Therefore estimates of R_0 and the effect policy decisions have on R_0 are central to community response to an infectious disease [26, 27, 5].

What is less understood is how to estimate this from data and from models. In demography this could be estimated by taking the total number of offspring in a year and dividing this by the total adult population to arrive at the average offspring per adult produced in a year. However, there are clear issues with this including how to draw a line between adults and offspring and how to adjust for variations within a year. The other difficulty with this method is that it is not at all predictive. It may describe whether your population was increasing or decreasing *last* year, but it does not predict whether the population will grow in the following year. This is a particular deficiency when considering infectious diseases and epidemiology since the goal is to develop strategies to counter the spread of diseases including ones that are currently spreading. Having some predictive model helps guide policy decisions during disease outbreaks.

The concept of the basic reproduction number was actually described in the 1920's in the context of demographics about 60 years before it was well established in the study of the spread of diseases. In demographics R_0 is a measure of the number of offsprings per each female member of the population. Again, it is intuitive that larger R_0 implies faster growth. Demographics and epidemiology share many characteristics and many of the earliest research in both fields was performed by Lotka who was interested in population dynamics in a general sense.

There are two groups that dominate the development of modern epidemiology and provide an interesting insight into the role a quantitative theory plays in the development of science [61]. In the 1890's through the early 1900's Ronald Ross studied malaria and was the first to show that mosquitos transmit the malaria parasite. Although it had already been shown that certain parasites can live in the mosquito gut, Ross argued that the spread of malaria was caused by the mosquito as a disease vector and that controlling the mosquito population could prevent the spread of the disease within human populations. This idea was already "in the air" in the early 1900's with several practical trials of mosquito control to reduce the spread of dengue fever; however, these empirical methods often failed inexplicably. Ross was motivated to use quantitative relationships between insect control and disease spread to develop estimates for the amount of mosquito population needed to be reduced and how large an area needed to be treated. Ross developed mathematical models that were combined with data that indicated that it was not necessary to eradicate the entire mosquito population. Instead he estimated a threshold density below which the disease would not spread.

This work was extended and refined by George Macdonald in the 1950's. Macdonald extended Ross' theoretical model that had compartments for susceptible and infected humans and mosquito density, by including compartments that represented infected mosquitos and details about the mosquito life-stages and infectivity. Macdonald developed his theory at the same time that DDT was created and widely used. This lead Macdonald to focus on important aspects of control which extended the application of the theory. Additionally, Macdonald worked closely with field studies to refine estimates of key parameters. Together, the Ross-Macdonald provided a template, generally specific to the spread of malaria, for the interplay between theory and application.

In 1927 Kermack and McKendrick published one of the classic papers in mathematical biology – "A Contribution to the Mathematical Theory of Epidemics" [38]. This paper generalized the Ross-Macdonald theory and presented a very general theory of epidemics. A special case developed in this paper is the version of epidemiological models that most students learn and the one presented below. It is one of the most fundamental models used in biology sharing overlap with ecological, demographic and population models.

Although Kermack and McKendrick are credited with modern epidemiological models, it must be noted that Ross, Macdonald and Kermack, and McKendrick all focused on a population threshold that determines whether the disease spreads or not if it is introduced into a population. In some cases (as we will show below) the threshold and the basic reproduction number provide the same information. But in general, R_0 provides a more robust understanding about the spread of an infectious disease. It took almost 30 years for R_0 to replace the population threshold in epidemiology even though it was in widespread use in demography to estimate population decline. Epidemiology rests on incremental advancement in the science from observations that indicate the most important biological processes, for example the insect vector, to quantitative estimates based on the current knowledge that are compared to observations to determine next generation models. The theory also needed cross-fertilization to develop practical methods to combat the spread of infectious diseases.

There are many methods to calculate R_0. One of the most intuitive is to determine parameters for which the infected population increases, $\frac{dI}{dt} > 0$ when the disease is introduced into a susceptible population. This is equivalent to determining whether the disease free state is stable or not. We note that this is not the only method used to estimate the propagation of the disease and is sometimes difficult to provide biological interpretations.

6.2 Two Compartment Models

6.2.1 Model

Epidemiological models are excellent examples of *compartmental* models. The dependent variables in these models are conceptualized as compartments with connections between compartments representing transitions between compartments. An example compartmental diagram is shown in Figure 6.1.

Kermack and Mckendrick

McKendrick was a medical doctor as well as a competent mathematician who published more than 50 papers on epidemiological theory. Kermack was trained in mathematics and chemistry but an accident left him blind so he focused on theoretical studies. The 1927 paper written by Kermack and McKendrick is a relatively controversial paper but not because of its content. This paper is often cited as the foundation of the SIR model; however, the differential equations as described here are only described as a special case of a more general framework. In fact, the differential equations were explored by Ross, McDonald, and others. Kermack and McKendrick made important contributions – mainly by connecting the concept of per capita infection with a propagation threshold.

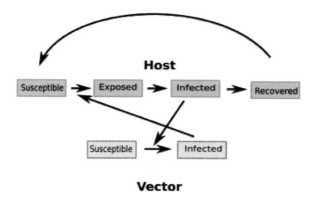

Vector

Figure 6.1: Schematic of a compartmental model. This example includes susceptible, exposed, infected, and recovered host populations with susceptible and infected compartments for the disease vector. This schematic is relevant for diseases such as malaria, but not for diseases spread person to person such as the flu.

The basic ingredients to epidemiological models require some distinction between members of the population in order to describe how the disease propagates. The simplest model distinguishes members who are sick and those that are not sick but have the potential to become sick. In the simplest diseases, the illness is passed from infected individuals directly to susceptible members. We will begin with an illness that never kills the infected individual but the individual never recovers from. In this case our population has two compartments shown in Figure 6.2 – susceptible, S, and infected, I. A word equation that describes the dynamics is,

$$\text{Rate of Change of } S \ = \ \text{Loss of } S \text{ to } I,$$
$$\text{Rate of Change of } I \ = \ \text{Gain of } I \text{ from } S.$$

If we denote the rate at which susceptible individuals become infected as k, we can make some logical inferences about this rate.

Host

Figure 6.2: Disease schematic with only two compartments.

First it must depend on both S and I since there should be no transfer from susceptible to infected if either of these compartments is zero, so $k = k(S,I)$ and both $k(0,I) = 0$ and $k(S,0) = 0$. There are many forms of this rate but if we assume that the interactions between susceptible and infected individuals is random we can assume that the law of mass action approximates the rate of interaction and that transmission of the disease proportional to the product of S and I, $k(S,I) = kSI$. We note that this assumptions is definitely not universally true – during the COVID pandemic it became clear that front line workers (e.g. doctors and nurses) necessary workers have higher rates of contracting the disease implying that the transmission was not random. Another example where random interactions is not appropriate are diseases that pass from mother to child. This is an assumption that we should exam closely if we find the predictions from our analysis are unreasonable. Realistically, there are no important diseases which can be represented with this model. These are diseases for which there is no cure, but is never deadly. Instead, we can think of this as a step-wise model that provides a possible road-map to understanding the dynamics.

6.2.2 Analysis

Given our assumptions, we can translate the word equations into differential equations,

$$\frac{dS}{dt} = -kSI, \tag{6.1}$$

$$\frac{dI}{dt} = kSI. \tag{6.2}$$

Recall that we assumed the population was constant since there were not births or deaths included – another way to see that this assumption is consistent with this model is to add these two equations to find,

$$\frac{dS}{dt} + \frac{dI}{dt} = 0,$$

$$\frac{d(S+I)}{dt} = 0,$$

$$S+I = N.$$

Where we used the fact that the sum of derivatives is the derivative of the sum and that if the derivative of a function is zero, then the function is constant (denoted N). We see that the total population is always constant and we can use this to simplify the equations. Since $S + I = N$, $S = N - I$ and we can substitute this into Equation 6.2,

$$\frac{dI}{dt} = kI(N - I).$$

This should be familiar since it is the logistic equation. We see that $I = 0$ and $I = N$ (that is $S = 0$) are the only steady-states and that $I = 0$ is unstable for all values of k. We can use the stability of the disease free state to understand how the disease spreads when initially introduced into the population. Because,

$$\frac{d\left(kI(N - I)\right)}{dt}\Big|_{I=0} = kN$$
$$> 0, \tag{6.3}$$

the disease free state is always unstable. In this overly simplified model, there is no parameter choice where the disease does not propagate, so the concept of a particular parameter value that separates the spread and decline of the disease state is not really applicable. Instead, all parameters lead to an increase in the infected compartment.

6.2.3 Sensitivity Analysis: Spider Plot

There are many measures of the spread of a disease. For epidemics, one measure might be the maximum of the infected population since this gives some indication of the strain on hospitals. For this two-compartment model, we know that the maximum is N so there would be no variation in the maximum value. Therefore all parameters including the total population, N, transmission rate, k, and initial condition would have the same sensitivity.

It also might be of interest to see how fast the disease is spreading. This could be quantified early in the disease process when the disease is first introduced or at different time points to see if the disease is increasing the rate of spread or not. This could be determined by the right-hand-side of Equation 6.3.

Another marker that provides information about the disease progression is the time the infected population reaches half of the maximum (N). This is a standard measure for sigmoidal curves referred to as the half-saturation constant . The workflow for this would be,

1. Select a parameter set

2. Solve the differential equation

3. Determine the time that $I = \frac{N}{2}$, $QoI = T_{half}$

4. Save this value.

We should note that for this example we could solve the equation analytically and find the half-saturation constant from the solution. We are using this as an example to motivate the sensitivity method. We will use a graphical method that provides information about the impact of parameter variations in a relative way. These methods are often referred to as "ranking" methods since the goal is to determine which parameters have more impact than others rather than quantify the effect of varying parameters. Spider plots are relatively simple to understand. We change each parameter, one at a time, sweeping through percent changes (above and below the nominal value). Plotting the QoI's simultaneously provides a quick way to judge the relative importance taking into account the percent differences.

We will add some more interesting behavior to the equations that we will solve numerically since the SI model is completely solvable, so is much less interesting. For demonstration purposes, we can add vital dynamics. We will maintain the two classes, but assume that there is an immigration term (or a birth term) that enters the susceptible population and a death term for the infected population. The new model is,

$$\frac{dS}{dt} = rS(S - \kappa) - kSI, \tag{6.4}$$

$$\frac{dI}{dt} = kSI - \delta I. \tag{6.5}$$

This gives us five parameters to consider: r, κ, k, and δ and the initial populations. To decide on a QoI, we first solve the equations for specific parameters. We will use qualitative parameters (e.g. not tuned

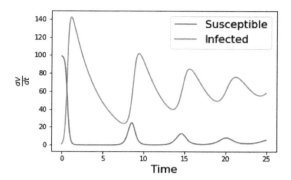

Figure 6.3: Susceptible/Infected dynamics including vital dynamics.

for any specific observations): $r = .05$, $\kappa = 100$, $k = .05$, and $\delta = .3$.
We will start with initial conditions, $S_0 = 99$ and $I_0 = 1$ so that the initial infected population is 1% of the initial populations. An example of the dynamics is shown in Figure 6.3. We see that adding vital dynamics can add oscillations, although this is parameter dependent.

We will use the ratio of I and the total population as the QoI – since this may oscillate, we will take the long-time solution. it can be shown that there is a stable equilibrium, so this is reasonable choice. We show the spider plots for a 50% variation above and below the nominal parameter set (see Figure 6.4).

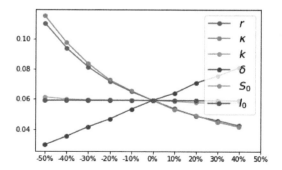

Figure 6.4: Spider plots comparing one-at-a-time variation of parameters in 10% increments. The QoI is $\frac{S_{steady}}{S_{steady} + I_{steady}}$.

Host

Figure 6.5: Disease schematic susceptible, infected, and recovery states.

We see immediately that r and κ lead to decreasing QoI while δ is positively correlated. Each of these has a stronger effect than varying any of the other parameters. In fact, it is impossible to see both the effect of varying S_0 and I_0 because they have exactly the same effect at all levels of variation.

6.3 Classical SIR

6.3.1 Model

We now move to the classic SIR (susceptible, infected, recovered) model represented by the schematic diagram in Figure 6.5. This model can be used to represent the dynamics of a disease which is not deadly but everyone eventually recovers. One could imagine a very mild cold, for example. Although we should point out that there are segments of the population that could be challenged by almost *any* immune system challenge. Immune compromised individuals, elderly, or other subsets of the population need to be accounted for differently. We are also still neglecting births, so we have to keep this in mind when we examine our results.

If we assume that the infection has a fixed recovery rate, γ, we can write the equations governing the disease dynamics,

$$\frac{dS}{dt} = -kSI, \tag{6.6}$$

$$\frac{dI}{dt} = kSI - \gamma I, \tag{6.7}$$

$$\frac{dR}{dt} = \gamma I. \tag{6.8}$$

It is worth noting that the road-map laid out by the SI model provides some insight into the next steps. If we add the equations we find that,

$$\frac{dS}{dt} + \frac{dI}{dt} + \frac{dR}{dt} = 0,$$

$$\frac{d(S+I+R)}{dt} = 0,$$

$$S+I+R = N.$$

So we can eliminate one of the compartments by replacing it with the difference between the total population and each of the other compartments – for example $R = N - S - I$. Because there are no births and deaths the total population is always N. For the SIR model, the population flows into the recovered compartment, which is not connected to the rest of the compartments in any other way so we can focus on the S and I dynamics and use the algebraic relationship, $R = N - S - I$, for R.

$$\frac{dS}{dt} = -kSI, \tag{6.9}$$

$$\frac{dI}{dt} = kSI - \gamma I, \tag{6.10}$$

$$R = N - S - I. \tag{6.11}$$

6.3.2 Analysis

Notice that for a steady-state either $I = 0$. This is can be seen in Figure 6.6, where the rates of infection and recovery are different. We see different dynamics for the susceptible population. in either case, $S = 0$ and $R = N$ as $t \to \infty$. In the first case, $k = 0.005$ and $\gamma = 1$. There is a wave of infection that recedes. The second case has a faster recovery time, $\gamma = 2$ and we still see a wave that passes. The second case has a residual of susceptible people who have not had the disease.

This disease wave that sweeps through the population is relatively simple here because the recovered population is immune and there are not variations in the susceptible population. Still, the peak of the wave

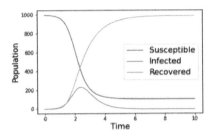 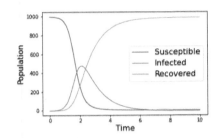

Figure 6.6: Dynamics for low infectivity, $N = 1000$, $k = 0.005$, and $\gamma = 0.5$. Comparison between two different disease examples. One is less infective ($k = 0.005$) than the other ($k = 0.01$). These differences result in very different outcomes.

depends on the characteristics of the disease – How easily is it transmitted? How fast is the recovery time? For many epidemics the goal is to reduce the peak infected population to reduce the impact on the infrastructure – this leads to methods to "flatten the curve".

There are several other aspects of the dynamics that are worth noting. The peak for the higher infectivity disease occurs earlier and is higher. This has policy implications since this means the number of people who have the disease and may require treatment is higher and more concentrated. This had a tremendous impact on the hospital infrastructure during the waves of COVID and was the focus of a lot of the modeling to predict resource needs.

These observations can be used to develop QoI's that are geared to answering specific policy questions such as the effectiveness of disease prevention – for example social distancing in terms of COVID or condom use for sexually transmitted diseases.

We can also try and understand the estimates of R_0. For the SIR model, one estimate for R_0 can be seen by determining when $\frac{dI}{dt} > 0$. Going back to Equation 6.8, we can see that if $kSI - \gamma I > 0$, the infection grows. At the beginning of an epidemic most of the population is susceptible so that $S \approx N$ so,

$$kSI - \gamma I \approx I(kN - \gamma) \quad > \quad 0$$
$$\text{if}$$
$$\frac{kN}{\gamma} \quad > \quad 1.$$

We can define $R_0 = \frac{kN}{\gamma}$ as the ratio of the fraction of the population that is infected and the fraction that recovers. We can check

R naught (R_0)

R_0 is notation for the basic reproductive ratio and is a measure of whether a disease can spread throughout a population or not. Intuitively, if an infected individual is not able to infect at least one other person the disease will not spread since the secondary infections are fewer than the primary infections. In reality, this is far too simplistic of an idea. First of all this does not provide an algorithm for approximating the number of secondary infections. Second, diseases can be far more complicated with far different dynamics in the event of reinfection, for different members of the population and different behavior patterns. Therefore there has been widespread interest in how to calculate R_0 more accurately.

The most straightforward definition of R_0 is through the rate of change of the infected population. If this is positive, the disease is expanding and R_0 is large. Otherwise R_0 is small. The difficulty with this is that it is difficult to interpret this biologically in any general sense. It is the method that we rely on here, however. Another method is the survival function approach [27] where there is a large population and $F(a)$ defines the probability that an individual remains infectious for a time interval of length a. The average number of secondary infections caused by an infected individual will be denoted $b(a)$. Then, we can define $R_0 = \int_0^\infty b(a)F(a)da$. This can be extended to complex models but rapidly becomes unwieldy.

The most general method is referred to as the "next generation matrix". This matrix is a statement of how the susceptible and infected populations change over one generation. This matrix defines the evolution of the population in discrete steps. The maximum eigenvalue of this matrix defines R_0. This is arguably the most general and accurate method to use. However, it is quite intricate in general so we will not focus on it.

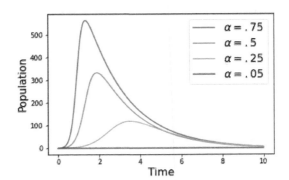

Figure 6.7: Comparison of the infected population for different values of α.

whether this makes sense for our examples. The simulations shown in Figure 6.6 have $R_0 = \frac{0.005 \times 1000}{0.5} = 10$ and $R_0 = \frac{0.01 \times 1000}{0.5} = 20$, so the disease propagates. If we remove a portion of the population so that the susceptible population that has the potential of contracting the disease is αN, we can vary α and determine what percentage of the population needs to be quarantined to prevent the spread of the disease. Examining Figure 6.7, we find that as long as $\alpha < .05$ the disease essentially does not spread – that is an estimate of about 95% of the population must be removed from the susceptible population.

6.3.3 Sensitivity Analysis: Tornado Plots

We will again make the model a bit more complex so that there are a few more parameters to deal with. Just as in the SI model, we can include basic vital dynamics so that there are births of susceptible and deaths of infected. Our new model is,

$$\frac{dS}{dt} = rS(\kappa - S) - kSI, \tag{6.12}$$

$$\frac{dI}{dt} = kSI - \gamma I - \delta I, \tag{6.13}$$

$$\frac{dR}{dt} = \gamma I. \tag{6.14}$$

Our parameters are the growth rate ($r = .1$), the carrying capacity ($\kappa = 2 * N$), infectivity rate ($k = .01$), death rate ($\delta = .01$), recovery rate ($\gamma = .5$) and total initial population ($N = 1000$).

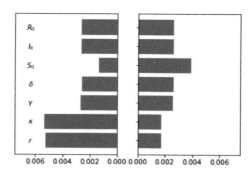

Figure 6.8: Tornado plot of the effect of a 50% increase or decrease in each parameter. We see that there is some asymmetry implying that decreasing κ and r has a stronger effect on the QoI than increasing these parameters. Also, this has a stronger, relative affect on the QoI.

To consider the sensitivity, we can compare the QoI for the low, versus high values of the parameters. The specific QoI shown in is the same as the one for the SI model: $\frac{S_{steady}}{S_{steady}+I_{steady}}$. One way to visualize this is using tornado plots. These are horizontal bar plots that show differences in increasing (on the right) and decreasing (on the left) each parameter, one at a time. Just as spider plots give a visual and relative comparison, tornado plots quickly show whether increasing or decreasing a parameter has a larger effect. It also compares the variation in each parameter. A plot for this example is shown in Figure 6.8.

There are several caveats – this is still one-at-a-time sensitivity. A slightly more subtle issue is that just looking at the increase and decrease assumes that the behavior is linear. If the sensitivity is nonlinear (for example quadratic) it is possible to mis-represent the variation.

6.4   Waning Antigens

There are numerous examples of infectious diseases where the SIR model is not accurate because recovery does not grant permanent immunity as in diseases like measles. Instead, after a waiting time a

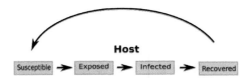

Figure 6.9: Disease schematic with four compartments.

recovered individual becomes susceptible so the model becomes an SIRS. It is important to note that the rate of transition between recovered and susceptible matters a lot as it modulates the speed of disease propagation. It is also a key measurable quantity used to estimate vaccination strategies. The goal of vaccination is to move members of the population from the susceptible category into the recovered category. There are many assumptions underlying this strategy that have varying degrees of correctness and depend on specific diseases. For example, we are assuming that recovery from the disease is equal to immunity from vaccination which is a tenuous assumption without proof. There are many diseases where there is a substantial difference between an induced immune response from a vaccination and the response from the active infection. However, all vaccine strategies aim to decrease R_0 by reducing the number of susceptible individuals that an infected individual to interact with. If a vaccine program is slower than the timescale of recovery to increased susceptibility, R_0 may never decrease enough to stop the spread of the disease.

For now, we will assume that there is no vaccine for the disease and that once an individual has been infected and recovered they move from the recovered compartment to the susceptible compartment at fixed rate (see Figure 6.9). We are again neglecting vital dynamics so we expect the total number of individuals in the population to remain constant – which will guide our analysis.

6.4.1 Model: SIRS

Using the familiar compartmental model, we describe the dynamics of all compartments. The equations are the same as the SIR model except

that we have added the transition from recovered to susceptible at a
rate α.

$$\frac{dS}{dt} = -kSI + \alpha R, \tag{6.15}$$

$$\frac{dI}{dt} = kSI - \gamma I, \tag{6.16}$$

$$\frac{dR}{dt} = \gamma I - \alpha R. \tag{6.17}$$

Again, we have a conservation law since there are no deaths in this
model: $\frac{dS}{dt} + \frac{dI}{dt} + \frac{dR}{dt} = 0$ so that $R = N - S - I$ and the system of three
Equations 6.16 - 6.17 can be reduced to two equations,

$$\frac{dS}{dt} = -kSI + \alpha(N - S - I), \tag{6.18}$$

$$\frac{dI}{dt} = kSI - \gamma I. \tag{6.19}$$

Interestingly, the estimate for R_0 we used for the SIR model, based
solely on whether I grows or decays, is the same for the SIRS model.
This suggests that this simple definition of R_0 is missing something
– either that or our understanding of the spread of a disease is not
sophisticated enough. The models are not the same, the behavior is
also not the same so it seems hard to imagine our measure of disease
propagation should be the same.

We can compare among different parameters and see important differences. In Figure 6.10, we show dynamics where the only parameter
difference is the rate of antigen waning – that is how quickly recovered individuals enter the susceptible population. For low values, we
see one isolated peak and as the rate of transition between recovered
and susceptible increases we see repeated waves of infection and eventually a state where the infection never recedes and is endemic in the
population. We will focus on the mathematical mechanisms for the
waves of disease and consider how sensitivity analysis can provide insight into controlling secondary waves rather than digging deeper into
the epidemiological aspect of R_0.

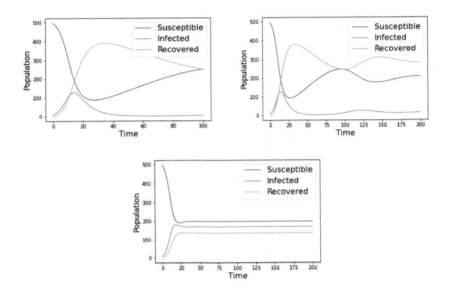

Figure 6.10: Comparison between three different disease examples where there is waning antigen so that recovered individuals can become susceptible. The parameters are: $k = 0.001$, $\gamma = .2$, $S_0 = 500$, $I_0 = 10$, $R_0 = 0$. The transition from recovered to susceptible varies from $a = 0 : 01, \alpha = 0.0125$ and $\alpha = 0 : 25$.

6.4.2 Analysis

How do we understand the differences in Figure 6.10? The first step is to return to our standard starting point: Steady-state analysis. Looking at Equations 6.19 and 6.19 to find steady-states, we can see that $I = 0$, $S = N$, and $R = 0$ is a steady-state. But there is another one with $S = \frac{\gamma}{k}$ and $I = \frac{\alpha(N - \frac{\gamma}{k})}{\alpha + \gamma}$, $R = N - \frac{\alpha(N - \frac{\gamma}{k})}{\alpha + \gamma}$. Linearization provides the reasons for the repeated waves as well as indication of the change between endemic and transient (see homework 6.8).

6.4.3 Sensitivity Analysis: Cobweb Diagrams

What can sensitivity analysis tell us about the behavior of the disease? Since the generic behavior – meaning the behavior that depends on a range of parameter values rather than one specific set of parameter values – is a decaying oscillation, it might be useful to know what controls the number of waves in a given time period. It also might be useful to know how the peaks of the waves depends on the parameters. This gives two QoIs that are easier to say in words than in equations. One way would be to determine the number of times the derivative of $I(t)$ is zero in a given time interval. Since this happens at the peaks and troughs it indicates the number of times the infectivity rate changes direction. We could also determine the value of I at these points to determine the amplitude of the oscillations. This type of QoI illustrates a key difference between sensitivity analysis that is useful in biological settings than other, more engineering-based, settings. In engineering applications it is often more standard to have the dependent variables of a model be identified as the QoI. Often this is because engineering models are often developed with specific reliability issues in mind.

Oscillations are not simple to measure, numerically – especially when the shape of the curves can change. We will use a QoI that distinguishes between endemic and transient – namely, the value of the infected population is small or not (that is whether the infection is endemic or not). To indicate the sensitivity, we will use cobweb plots. The idea is to choose parameter sets where each parameter is perturbed from the basal state. The perturbations are randomly chosen and independent and are typically a percentage change from nominal. Once the parameter set is chosen, the solution is determined numerically. The QoI is sorted into bins. We are using two bins although others can be used. The parameters that lead to specific bins are connected visually. There are many different coloring options that help identify different aspects of the sensitivity. In Figure 6.11, we show a simple coloring identifying which parameters lead to high/low infected population.

There are a two parameters that appear to be well separated. When the infection tends to pass and not be endemic α is lower and γ is higher. There are other inferences that can be drawn from the shading – for example, γ is very certainly a parameter that moves the system between endemic and not while the initial conditions do not matter much at all.

We should also note that cobwebbing is the only global method that we have discussed in this chapter – the others are one-at-a-time. Global methods carry much more information, but are typically more complex to code up.

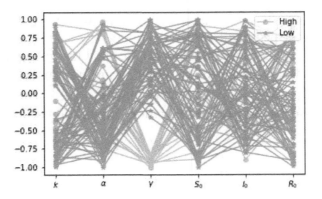

Figure 6.11: A variation of 100% indicating parameters that distinguish between high and low.

6.5 Caveats and State of the Art

We should make some mention about the scaling that we are using. There are two standard ways to implement the SIR-type models. These can be either dimensionalized or nondimensionalized – there are useful reasons for each, but it is important to understand the differences.

We can start with the dimensionalized version,

$$\frac{dS}{dt} = -kSI + \alpha R, \tag{6.20}$$

$$\frac{dI}{dt} = kSI - \gamma I, \tag{6.21}$$

$$\frac{dR}{dt} = \gamma I - \alpha R. \tag{6.22}$$

All the state (dependent) variables are measured in units of population (individuals, or appropriate units of individuals such as millions). We can measure the independent variable in units of time (for epidemics days is often appropriate). The units of the left-hand-sides must match the units on the right-hand-sides. Therefore

$$\left[\frac{dS}{dt}\right] = \frac{\text{population}}{\times}\text{time} = [kSI] = [k][S][I].$$

Therefore, $[k] = \frac{1}{\text{population time}}$ and k is a per capita rate. At the same time, α an γ are rates, $[*] = \frac{1}{\text{time}}$. This is important when using parameters relevant to specific diseases.

As before, the conservation law implies that the total population is N. This provides a scaling for the populations in terms of fractions of the total. Define $s(t) = \frac{S(t)}{N}$, $i(t) = \frac{I(t)}{N}$, an $r(t) = \frac{R(t)}{N}$. Then $\frac{dS}{dt} = N\frac{ds}{dt}$ and $[s] = 1$. We can re-write the equations in non-dimensional form,

$$\frac{ds}{dt} = -Nksi + \alpha r, \tag{6.23}$$

$$\frac{di}{dt} = Nksi - \gamma i, \tag{6.24}$$

$$\frac{dr}{dt} = \gamma i - \alpha r. \tag{6.25}$$

The parameter group, Nk has dimensions of $\frac{1}{\text{time}}$ since,

$$[Nk] = \frac{[k]}{[N]},$$

$$= \frac{\text{population}}{\text{population} \times \text{time}},$$

$$= \frac{1}{\text{time}}.$$

This means that the initial conditions for the two scalings must be consistent (non-dimensional scales are on the order of one, dimensional scale on the order of N). The parameters that involve nonlinear combinations of variables need to be scaled to either per capita or per time.

6.6 Problems

Problems 6.1 *Consider the SIRS model,*

$$\frac{dS}{dt} = -kSI + \alpha R, \tag{6.26}$$

$$\frac{dI}{dt} = kSI - \gamma I, \tag{6.27}$$

$$\frac{dR}{dt} = \gamma I - \alpha R. \tag{6.28}$$

(a) *Find the equilibria.*

(b) *Compare these with the two-variable version using the conservation of the population (Equations 6.16, 6.17, and $S + I + R = N$).*

(c) *Use any method to determine the stability of these equilibria.*

(d) *Demonstrate the stability properties numerically using specific values of parameters that imply stability for the steady-states.*

Problems 6.2 (a) *Show that the maximum of the two-compartment (SI) model is N*

(b) *Demonstrate this using the numerical codes.*

(c) *What do you note about the approach to the equilibria?*

Problems 6.3 (a) *Solve Equation 6.3 for $I(t)$*

(b) *Write an expression for the analytic value of the half-saturation constant.*

(c) *Use this to compare with the sensitivity method approximated numerically.*

(d) *Change the QoI to the value of the rate of change of the infected compartment evaluated at an increasing sequence of points, $(0, t_1, t_2, t_3, t_4)$. Compare the sensitivity plots for different times. What do you see as t_4 gets large?*

Problems 6.4 *What do the sensitivity methods show for the sensitivity of the maximum of the two compartment model with respect to parameters? Are there differences in the methods?*

Problems 6.5 *Suppose the population is split into two different subsets: healthy and immune compromised.*

(a) *Sketch a schematic where there are 4 compartments: Healthy susceptible, compromised susceptible, infected, and recovered.*

(b) *Write a model that expands the susceptible compartment S in the SIR model to $S = S_h + S_c$.*

(c) What happens if S_c do not become infected but are removed from the population? Show this numerically.

(d) Is this realistic? What happens to the total population? What would happen if there was a transition between S_h and S_c?

Problems 6.6 *Consider the SEIR model for these cases:*

- *Including a quarantine for sick people.*

- *Include vaccination.*

- *Include age structured: The simplest way to do this is to consider SEIR models for some division of the population: Youth, Adults, Aged. Youth become Adults at a particular rate. We then have compartments for susceptible, exposed, infected, recovered for each of the groups: youth, adults, and aged. Just as susceptible can become exposed, youth can become adults at a specific rate.*

For all of these, provide representative simulations of the dynamics for some simple parameters. Consider the steady-state behavior. Can you find an analytic representation of the steady-states?

Problems 6.7 *Explore different methods to quantify the number of oscillations occur in a fixed interval. You might explore the command* `tspan[np.argmax(np.gradient(yp.y[1],tspan)<0)]` *as a starting place. Here* `np.gradient(yp.y[1],tspan)` *is the numerical derivative of* `yp_y[1]` *and* `np.argmax(a<0)` *provides the index of a that is the first occurrance of a negative value of a. The time this happens is in* `tspan`*.*

Problems 6.8 *Use sensitivity to determine the parameters that are mainly responsible for distinguishing between endemic and transient disease dynamics.*

7

Microbiology

Quantitative models of microbiological processes have had a direct impact on disease treatment as well as optimizing engineering processes that rely on healthy microbes. We focus on examples of specialized experimental techniques and general bacterial phenotypes. Sensitivity methods include correlation, ranked, and partial ranked correlation coefficients.

7.1 Historical Background

Microbiology refers to the study of organisms at a very small scale ranging from nanometers to microns including viruses, fungi, and bacteria. We will concentrate on bacterial dynamics for several reasons. First because the dynamic behavior is quite rich and offers a connection between a wide range of scientific disciplines including biophysics, ecology, epidemiology, and population dynamics. Second, the science of bacterial microbiology has been one of the true successes in quantitative biology that is tempered by unique challenges. The rate at which the world moved from very little understanding of the cause of diseases to the development of antibiotics has been astounding; however, we are also confronted with one of the most serious challenges in health care in the coming decades. Antibiotic resistance is becoming far more prevalent and we appear to be on the cusp of revisiting a time before antibiotics were effective.

The study of microbiology is intimately connected with ecology, epidemiology, and population dynamics. We have already seen how ecological perspectives have substantial overlap with immunological perspectives. This overlap can be seen when studying disease-causing microbes since they are the source of the disease so directly effect the

DOI: 10.1201/9781003316930-7

population of disease carriers. It is also important to note that microbiology can be viewed through the lens of ecology focusing on the
population of microbes and their dynamics. So what distinguishes microbes from other organisms? In other words, why separate this topic
from ecology? This question has multiple answers that help define the
field and explain why mathematics plays an important role in microbiology. One answer is that microbiology requires specialized tools to
study the organisms. The study of the very small is intwined with the
engineering needed to produce microscopes. Single lens microscopes
are not capable of focusing on the micrometer scale needed to see bacteria and protozoa due to chromatic aberration. At these scales small
variations in the foci of a single lens microscope lead to blurring as
different wavelengths of light are collected into slightly different areas.
By taking one lens with a short focal length to collect the image and a
separate lens to enlarge this, extremely detailed images can be seen.

Antonie van Leeuwenhoek is often credited with developing the
first compound microscope in the 1600's. This is not strictly true
since compound microscopes were developed by Hans and Zacharias
Janssen 50 years previously. However, Leeuwenhoek introduced several key improvements and used this to study an extraordinary range of
material including the first accurate description of red blood cells, descriptions of capillaries, tissue and "very little animalcules". Because
of his thorough compilations of sketches and studies, he is often referred to as the *Father of Microbiology*. Interestingly, Leeuwenhoek
developed his microscope to study cloth that he manufactured and only
inadvertently noticed tiny organisms trapped in fibers. It was providen-
tial that he noticed these and was interested enough to pursue his investigations. His description of the life cycle of the flea helped promote
his microscope and lead to an explosion of information. But it would
be 200 years before Pasteur connected the "animalcules" to the cause
of disease.

A second reason to make a distinction between microbiology and
ecology, epidemiology, or other topics could be termed "plasticity".
Bacteria are far more flexible than other organisms. They have multiple respiration pathways and can move from aerobic to anaerobic respiration depending on the local environment. Thus there is no fixed
phenotype for a typical bacterium. Different genes can be up or down

regulated including genes that alter growth rate, induce the production of material to maintain cell walls and other structures, alter motility, form a biofilm, and hosts of other traits. This precludes true compartmental models at the population-level. Instead multiple compartments may need to be used to quantify the phenotypic expression.

Finally, bacteriology has undergone a dramatic paradigm shift within the past 50 years that has ramifications for health care, food safety, water quality, environmental science, agriculture, and many other applications that directly impact everyone. Prior to the 1980's the major view of bacteria had gone through important, but relatively incremental changes. In the 1800's the connection between bacterial types and diseases was demarcated including the identification of anthrax, leprosy bacillus, and plant associated diseases. This lead directly to treatments such as heat sterilization and vaccination strategies. As more understanding of bacteria was collected, examples of beneficial bacteria were also discovered including the process of nitrogen fixation within plant roots that drives photosynthesis.

In the early 1900's more and more bacteria were isolated and studied. Similarly connections between microbes and other organisms were discovered. It was found that certain insects could serve as the vectors for diseases and that viruses interact directly with bacteria although viruses were not formally distinguished from microbes until the 1920's. This distinction lead to the development of viral vaccines by Jonas Salk and others in the 1950's.

In the early 1980's systematic studies of aggregates of bacteria began to converge on a startling new concept – bacterial community dynamics are inherently different than individual dynamics. This, coupled with new evidence that bacteria in nature are typically found in biofilms. In fact, it is estimated that more than 98% of bacteria in nature live in these complex, structured communities. The impact of this shift is becoming more clear since many of the treatments developed for free bacteria are far less effective against bacteria within biofilms. More subtle, but equally important, differences in growth rates have been noted between free swimming bacteria and those in biofilm communities. Because antibiotic dosages are developed based on these measurements, it is very likely that how we deliver antibiotics is not optimal.

Mathematical modeling of bacteria to include all of this knowledge leads to very complicated models that include temporal and spatial dynamics (e.g. partial differential equations), physics, chemistry, and biology. In this chapter, we will focus on four bacterial models, using the pre-1980's paradigm of bacteria, that incorporate some of the broad development in microbiology but are still restricted enough to be readily studied. We begin with the development of a model bacterial growth in the case of a single nutrient. We then turn to a model of growth in a chemostat which is an important tool developed in the 1950's. We then describe two models of interacting phenotypes of bacteria

7.2   Bacterial Growth: Chemostat

Determining how to model bacterial growth is quite important. Often a logistic model is used to describe the growth of populations; however,

---

William Costerton (1934–2012)

The rapid change in our understanding of bacteria has happened within the last 40 years. One of the most influential scientists responsible for this shift was Bill Costerton. He authored more than 700 publications and was a tireless champion of studying bacteria in conjunction with the environment. He wanted to know how a bacterium knew the difference between a catheter and a river – which it became apparent they do. He pioneered the development of special equipment needed to study biofilms including the "rototorque" which was able to mimic the flow within a pipe but with much more control. His interest in quantification, modeling, experiment, and theory lead to rapid expansion in understanding the multiple bacterial phenotypes. He was the director of the Center for Biofilm Engineering which is a multidisciplinary center with a central aim of understanding bacterial biofilms by connecting engineers, microbiologists, mathematicians, and statisticians.

this requires some knowledge of the growth rate and carrying capacity. When developing more detailed models – for example a model that includes spatial variations – logistic growth may be overly restrictive. Instead it is useful to think about bacterial metabolism and how dynamic nutrient can be coupled with bacterial growth. In the 1940's Monod introduced an empirical model for the growth rate of a population of bacteria using one nutrient source [47]. Empirical derivations rely on experimental data to develop a functional relationship between nutrient concentration and growth rate. It is a relatively intuitive idea – beginning with a low concentration of nutrient, measure the average growth rate of a colony of bacteria. With each new experiment, incrementally increase the nutrient concentration and remeasure the growth rate. Monod observed that there was a monotonic relationship but that there was a relatively small concentration regime where there was an abrupt increase in growth rate. Monod found that a curve, referred to as a sigmoidal curve, was able to capture his experiments. Monod kinetics defines the growth rate, r, as a function of the nutrient concentration, N as $r(N) = \mu \frac{N}{K+N}$. Here μ is the maximum growth rate and K is the half saturation constant.

The same relationship can be derived using methods developed in enzyme kinetics. Bacteria rely on the cell wall to separate the interior cytosol from the exterior domain. In order to get nutrient from the external environment into the cell so that it can be used, channels in the wall must be used. A simplistic model of these channels assumes that specific proteins that are embedded in the cell wall interact with nutrient causing a conformation change allowing them to transport nutrient through the cell wall. Because there are a finite number of these protein transporters, we can treat the protein/nutrient system analogously with enzyme kinetics. If there is abundant nutrient, all the protein channels are occupied and adding more nutrient cannot drive more nutrient across the cell wall and thus there is a limit to the growth rate. At the same time, for low concentrations of nutrient the effect of doubling the concentration of nutrient will double the nutrient that can cross the cell wall – that is the relationship is linear at low concentrations. This leads to the same sigmoidal curve as Monod derived. In a general context, this is often referred to as Michaelis-Menton kinetics based on classical studies performed in a classic paper published in 1913. This is an example of a scientific misnomer since Michaelis-Menton kinetics

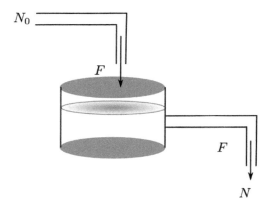

Figure 7.1: Schematic of chemostat with flow rate F, influent chemical c_0 and outflow concentration c.

is focused on enzyme chemistry and we are concerned with bacterial kinetics; however, the connection between the bacterially focused empirical work of Monod and the theoretically justified work of Michaelis and Menton does demonstrate the portability of mathematical models [35, 12].

7.2.1 Model

We now turn to a tool that has been traditionally used to culture bacteria – especially to maintain a specific metabolic rate that might impact experimental observations. For example, it is well known that antibiotics are more effective against rapidly dividing bacteria, so in the absence of nutrient, bacterial populations are essentially unaffected by certain classes of antibiotics. The chemostat is a tank connected to an inflow and an outflow (see Figure 7.1) where nutrients are pumped into the tank at a certain rate, bacteria within the tank consume nutrient and reproduce and the mixture flows out of the tank. Typically the rate of inflow and outflow are the same so the volume of fluid within the tank is maintained at a constant level. Multiple nutrients and multiple species of bacteria can be included, but we will concentrate on single species systems with a single, limiting nutrient.

The word equation relating to this system is:

Rate of Change of Variable $=$ What Comes In $-$ What Goes Out

For each variable, bacteria and nutrient, we have,

Rate of Change of Nutrient (N) $=$ Source From Inflow

$-$Consumption+Outflow

Rate of Change of Bacteria (B) $=$ Source From Growth $-$ Outflow.

The source of nutrient from the flow is related to the flow rate, F. We could assume a constant source, in which case we would have a constant proportional to the flow rate. We might have a bolus of nutrient added or periodic sources in which case the source would need to be adjusted. The outflow of both nutrient and bacteria is proportional to the outflow rate and must include the specific concentrations.

Finally, consumption of nutrient and growth of bacteria are related. Michaelis-Menton kinetics leads to a specific form for growth: $g(B,N) = \mu \frac{N}{K_N+N}B$. Consumption is typically proportional to this, $C(B,N) = \frac{1}{Y}g(B,N)$. The parameter Y is called the yield rate and indicates how efficiently the bacteria use the nutrient. You can think of this as related to how many units of nutrient get converted to units of bacteria. Putting all this together, we find a chemostat growth model as,

$$\frac{dN}{dt} = N_0F - \frac{1}{Y}\frac{\mu N}{K_N+N}B - FN \tag{7.1}$$

$$\frac{dB}{dt} = \frac{\mu N}{K_N+N}B - FB. \tag{7.2}$$

where N_0 is the influent concentration of nutrient. We also need initial conditions – we will use zero nutrient and constant bacteria so $N(0) = 0$ and $B(0) = B_0$.

7.2.2 Analysis

We can now follow our typical route of steady-state analysis. We look for constant solutions which must satisfy,

$$0 = N_0F - \frac{1}{Y}\frac{\mu \hat{N}}{K_N+\hat{N}}\hat{B} - F\hat{N} \tag{7.3}$$

$$0 = \frac{\mu \hat{N}}{K_N+\hat{N}}\hat{B} - F\hat{B}. \tag{7.4}$$

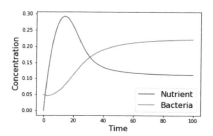
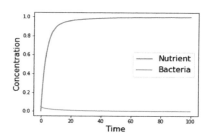

Figure 7.2: Dynamics for the bacteria/nutrient chemostat. We show two different outcomes. The parameters are $N_0 = 1$, $F = .05$, $Y = .25$, $\mu = .5$, $K_n = 1$ (coexistence) and $N_0 = 1$, $F = .3$, $Y = .25$, $\mu = .5$, $K_n = 1$ (washout)

From Equation 7.4, it is easy to see that either $\hat{B} = 0$ or $F = \frac{\mu\hat{N}}{K_N + \hat{N}}$. $\hat{B} = 0$ is the "wash-out" state where there are no bacteria in the chemostat. If there are no bacteria Equation 7.3 shows that $\hat{N} = N_0$, that is the nutrient equilibrates to the source concentration so there is one equilibrium $(N_0, 0)$.

The other equilibrium occurs by assuming that $B \neq 0$. In this case Equation 7.4 implies

$$
\begin{aligned}
F &= \frac{\mu\hat{N}}{K_N + \hat{N}}, \\
\hat{N} &= \frac{FK_N}{\mu - F}.
\end{aligned}
\tag{7.5}
$$

Using this in Equation 7.3 we find, $\hat{B} = Y(N_0 - \hat{N})$. We can determine the stability of these two equilibria – interestingly only one of the two equilibria is stable for any parameter set. By adjusting different parameters we see very different behavior. We plot the two cases in Figure 7.2.

7.2.3 Sensitivity Analysis: Correlation Coefficient, Pearson's Moment Correlation

One of the most useful ways to interpret sensitivity is an extension of ideas presented in Chapter 4, where we sampled over parameter space and considered how the scatter plots indicated correlation between parameter variations and QoI's. Recall that one way to quantify this is to use linear regression (i.e. approximate the point cloud with a linear equation and determine the slope). A more general way to think about this is to focus on *correlation* – that is the statistical relationship between two variables. In the context of sensitivity, we are interested in the correlation between the QoI and the input parameters.

There are many viewpoints that one can take to understanding correlation including geometric, statistical, and algebraic. One broadly useful outlook is to consider variance and covariance. Variance refers to a measure of spread from the mean within a set of numbers. Covariance revers to a measure of relationship between two different random variables. One might expect that sensitivity measures would be related to covariance since one could use this to measure how the QoI moves with respect to variations in the input parameters. While this is true, it is important to recognize that covariance alone may not be enough. There may be large covariance without a strong correlation between two random variables if one of the variables itself has a lot of variation. Another way to say this is that covariance has units and needs to be measured against something to assign a scale. In 1895, Pearson derived a formula where the covariance was scaled by the combination of the individual variances. We define the variance in a sample of values as $\{X_i\}$ as $\sum_i (X_i - \bar{X})^2$ where $\bar{X}$ is the mean of $\{X_i\}$. Then the covariance between two different samples, $\{X_i\}$ and $\{Y_i\}$ is defined as $\sum_i (X_i - \bar{X})(Y_i - \bar{Y})$ [54, 40].

We now want to generalize this to a relationship between a QoI that depends on multiple parameters. To be precise we will define the QoI as y and denote the relationship (e.g. the model) between y and m parameters, $\{p_1, p_2, ..., p_n\}$ as $y = F(p_1, p_2, ..., p_m)$. To consider how the QoI varies as the input parameters vary, we take n samples over the j parameters, generating n QoI's. We arrange all the sampled parameters into an $(n \times m)$ matrix P with entries p_{ij}. The rows of P are one parameter set and yield an output. Pearson's correlation coefficient needs to be defined for each of the m parameters – that is we are interested in the correlation between the QoI and each of the n parameters. We define the correlation coefficient as,

$$r_{p_j,y} = \frac{\sum_{i=1}^{n}(p_{ij} - \bar{p}_j)(y_i - \bar{y})}{\sqrt{\sum_{i=1}^{n}(p_{ij} - \bar{p}_j)^2 \sum_{i=1}^{n}(y_i - \bar{y})^2}}. \tag{7.6}$$

This number is always between 1 and -1. We can then compare the Pearson's coefficient to both rank the importance of each parameter and provide an interpretation for the magnitude.

Suppose our QoI for the chemostat problem is the total number of bacteria produced in a given time period with a fixed concentration of nutrient source. Mathematically, this can be written as,

$$y = \int_{t=t_1}^{t=t_2} B(t) \, dt.$$

We can then examine which parameter has the most impact on the bacterial production. There are two aspects that we have not focused on. In particular, the sign on r_i has meaning, just as the slope of the regression line does. If $r_1 > 0$, the QoI is positively correlated. This means increasing the parameter leads to an increase in the QoI. Similarly for negative r_i.

Secondly, we have to mention the sampling method. Here is a place where sampling matters a lot. In Chapter 4, we assumed that the parameters were distributed uniformly. That is it was equally probably that a parameter value might be chosen within its range. However, there are many situations where we do not expect uniformly distributed parameters. Perhaps normally distributed is more realistic. It is even possible that there is a reason to assume that the parameters are not distributed symmetrically. Regardless, we should sample in a way that balances this. There are two main ways to do this: Latin Hyper Cube Sampling (LHS) and Monte Carlo sampling.

The main idea between LHS is to divide the parameter space into regions where the probability of drawing from each region is equal. This has a nice visual interpretation where one can think of the probability distribution as a function (PDF). Normally distributed parameters have Gaussian PDF's and if there is reason to believe the parameters have different distributions one can adjust the PDF and must sample accordingly. Once the PDF is partitioned, sampling randomly from each partition leads to samples that reflect the underlying statistics.

Monte Carlo sampling is a method that can be used to draw samples from a PDF. We will not show any algorithms for this since most platforms including MATLAB, Python, and R have samplers that can be used to generate a random sequence of parameters based on the

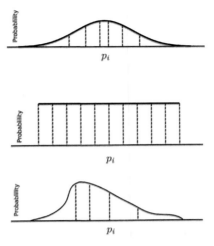

Figure 7.3: Comparison of equi-probable partitioning of the probability distribution for a single parameter. The upper panel is a normal distribution, where partitioning is finer near the mean (area under the curve for each partition is the same). The middle shows partitioning a uniformly distributed parameter into equal area (e.g. probability). The lower panel shows an asymmetric probability distribution.

underlying PDF. The basic idea is to draw a random parameter value within an interval. This value is kept as a sample if the value of the PDF at the picked value is less that the cumulative distribution of that point. This keeps samples more often from regions where the PDF is high.

Once we have this in place, we take m samples and calculate the correlation coefficient where $y = QoI(p_1, p_2, ..., p_n)$. How many samples are needed for the correlation to be valid? Unfortunately there is no fixed answer to this. Technically correlation coefficients can be calculated with at least two observations for each parameter. In practice, how many is enough depends on the model and complexity.

An example for the chemostat with

$$QoI = \int_{t=0}^{t=100} B(t)\, dt. \qquad (7.7)$$

Most examples here are not limited by the number of samples that can be taken, but if the model is very complex a smaller number of samples

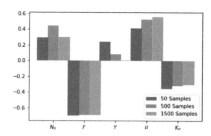

Figure 7.4: Comparison of the Pearson's sensitivity estimate for the parameters in the chemostat equations with different numbers of samples.

may be useful – but this must be done with care. Consider the Pearson correlation coefficients shown in Figure 7.4. There is not much variation in as the sample number increases, except for the sensitivity with respect to Y. How do we understand this? Recall that Pearson's is a linear-regressive measure and we need to examine the scatterplots. We show these in Figure 7.5 and see that Y has a lot of variation, but is essentially flat. Therefore under-sampling may introduce errors. Scatterplots also confirm that the signs of the correlation are correct.

7.3   Multiple State Models: Free/attached

One of the main criticisms to the standard chemostat model is that bacteria do not stay in suspension very well. Even in a chemostat that

Working with the computer codes: Steady-states

Looking at the equations, we can ask whether the washout state is stable or not. Looking at the code `Chemo_SS.py`, we find the washout steady-state of $(1,0)$ (at least with the given parameters – note that there are more than one. Looking at the steady-state, the Jacobian and corresponding eigenvalues we find that this equilibria is unstable (with eigenvalues -0.05 and 0.2). We can also check the dynamics near this state and verify the result.

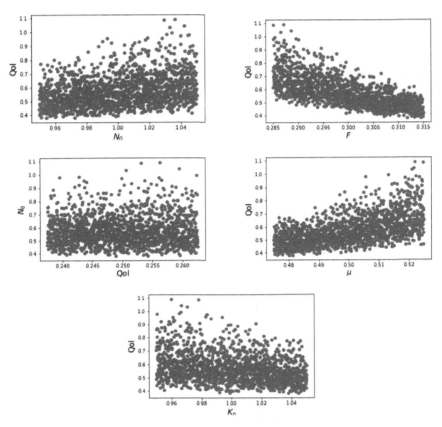

Figure 7.5: Scatterplots for each parameter against the QoI.

is "well-mixed", there are regions of the container where the flow is reduced and bacteria tend to collect there. Even more interestingly, as bacteria near a solid surface their motility is altered. In fact, there is accumulating evidence that bacteria actively seek to remain near a wall and can attach to the wall. As they attach there are multiple signaling pathways that are triggered inducing dramatic phenotypic alterations. One of the most important phenotypes controls the production of extracellular polymer which can provide a backbone for a biofilm to form. The term biofilm refers to an aggregate of microbes attached to a surface. This concept was not separated as a distinct process until the 1980's. Although groups of aggregated bacteria were observed several hundred years ago, the importance was not understood. Currently, microbiology is undergoing a paradigm shift where bacteria are recognized as interacting with each other and the environment in much more complicated ways than was understood previously. This has an impact on drug discovery, disease treatment, industrial applications and many other settings. Here we will only focus on the distinction that some bacteria are freely suspended and some are attached to a surface.

7.3.1   Model: Freter

To model the transition between bound and unbound bacteria can start simply by considering two different compartments where one compartment denotes the density of bound bacteria, B_b and the other is the unbound density, B_u. One needs to be careful about units, however. Density is typically defined as mass per volume, but this is not appropriate for bacteria that move from free suspension to being attached to a surface. Instead, the units of B_b and B_u are different which must be accounted for since attached bacteria are attached to a two dimensional surface while unattached bacteria move in three dimensions. Thus we define units for these as mass per area and mass per volume, respectively [4].

A simple model in a chemostat that neglects any process other than inflow/outflow and attachment/detachment can be written based on the system in Equations 7.1 and 7.2. We can assume that unattached bacteria are fed into the system with concentration B_0 and the attachment rate is α and detachment rate is a constant β. One should also consider whether bound bacteria are allowed to exit the system or not. In this example, we will not allow the bound bacteria to flow out of the system.

We will also define the area available for binding as A and the volume of the chemostat as V. The ratio of the volume and area provide the conversion so that the mass concentration of bacterial detaching (measured in mass per area) matches the units for the free bacterial density (mass per volume). The model becomes,

$$\frac{dB_u}{dt} = B_0 F - \alpha B_u + \frac{V}{A}\beta B_b - F B_u \tag{7.8}$$

$$\frac{dB_b}{dt} = \frac{A}{V}\alpha B_u - \beta B_b. \tag{7.9}$$

We can start with a clean chemostat so that $B_u(0) = B_b(0) = 0$.

There are two additional considerations that one might want to think about. First, the rate of bacteria attaching the wall might depend on the density of bacteria already attached to the wall. That is as the wall is covered by bacteria it may become more difficult for bacteria to find a location to attach. Second, what should happen to the progeny of bacteria that are attached to the wall? Do they all remain attached or do some of them detach? Does this depend on the density of wall attached bacteria?

To answer the first question, we can make several different assumptions, but assuming that attachment decays as the density of wall attached bacteria increases seems reasonable. Therefore α becomes a function of B_b. An example might be $\alpha = \frac{\alpha_{max}}{1+B_b}$ where α_{max} is the maximum attachment. The second question requires including growth of wall attached bacteria. We can assume that growth is related to the nutrient – for example, proportional $\frac{N}{K+N}$ as in Monod kinetics. Then there is a fraction of the progeny that remain attached, and the rest become detached. Here we will assume that the fraction is related to the density of wall attached bacteria (highlighted in the Equations 7.10–7.12). Our equations then become,

$$\frac{dN}{dt} = N_0 F - \frac{1}{Y}\frac{\mu N}{K_N+N}(B_u+B_b) - FN, \qquad (7.10)$$

$$\frac{dB_u}{dt} = B_0 F - \frac{\alpha_{max}}{K_\alpha+B_b}B_u + \frac{V}{A}\beta B_b - FB_u$$
$$+ \left(1 - \frac{B_b}{K_b+B_b}\right)\frac{1}{Y}\frac{\mu N}{K_N+N}B_b, \qquad (7.11)$$

$$\frac{dB_b}{dt} = \frac{A}{V}\frac{\alpha_{max}}{(K_\alpha+B_b)}B_u - \beta B_b$$
$$+ \frac{A}{V}\frac{B_b}{K_b+B_b}\frac{1}{Y}\frac{\mu N}{K_N+N}B_b. \qquad (7.12)$$

Clearly there are other choices that could be made that reflect different assumptions. But we can see a variety of behaviors here. Figure 7.6 shows different dynamics that occur as we adjust some of the parameters. It is useful to think about parameters that can be adjusted in different experiments. For example, it is easy to study how differences in N_0 and F affect the dynamics. These are easily changed – in

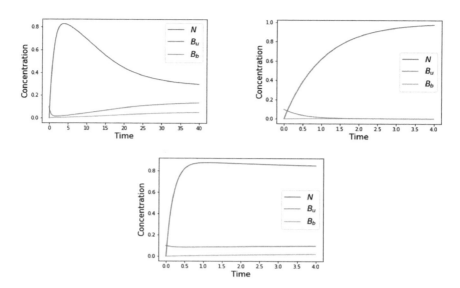

Figure 7.6: Dynamics for the bacteria/nutrient chemostat. We show two different outcomes. The parameters are $N_0 = 1$, $Y = .1$, $K_n = .5$, $F = 1$, $B_0 = 0$, $\alpha = .1$, $k_\alpha = .1$, $V = 10$, $A = 1$, $\beta = .2$, $K_b = .5$ for both examples and $\mu = 1$ (coexistence) $\mu = .1$ (washout). The bottom figure uses $\mu = 1$, $F = 5$ and demonstrates a fast transition to equilibrium.

fact, these can be changed dynamically throughout the experiment. But what about other parameters such as μ or K_b? These can be adjusted to some extent by changing the nutrient or genetically altering the bacterial strain – but this clearly alters the conclusions that can be made from the experiments.

7.3.2 Analysis

We can use the fast transition to simplify our system. For the parameters $N_0 = 1$, $Y = .1$, $K_n = .5$, $F = 5$, $B_0 = 0$, $\alpha = .1$, $k_\alpha = .1$, $V = 10$, $A = 1$, $\beta = .2$, $K_b = .5$, and $\mu = 1$, there is a transient where the nutrient rapidly approaches and equilibria. If we neglect this transient, $\frac{dN}{dt} \approx 0$, and we can determine the steady-state value of the nutrient, $\hat{N}$, in terms of the other variables,

$$N_0 F - \frac{1}{Y} \frac{\mu N}{K_N + N}(B_u + B_b) - FN.$$

We could solve this for N to find two values, one of which is physical. We can also see that this implies,

$$\frac{1}{Y}\frac{\mu N}{K_N + N} = \frac{N_0 F - FN}{B_u + B_b}.$$

This can be used in the equations for B_u and B_b,

$$\frac{dB_u}{dt} = B_0 F - \frac{\alpha_{max}}{K_\alpha + B_b}B_u + \frac{V}{A}\beta B_b - FB_u$$
$$+ (1 - \frac{B_b}{K_b + B_b})\frac{B_b}{B_u + B_b}(N_0 F - FN), \qquad (7.13)$$

$$\frac{dB_b}{dt} = \frac{A}{V}\frac{\alpha_{max}}{(K_\alpha + B_b)}B_u - \beta B_b$$
$$+ \frac{A}{V}\frac{B_b}{K_b + B_b}\frac{B_b}{B_u + B_b}(N_0 F - FN). \qquad (7.14)$$

There is still an N in both of these equations but we have saved a bit of algebra. After examining Equation 7.13, we can write this as a quadratic in N,

$$N^2 - \left((N_0 - K_N) - \frac{\mu}{FY}(B_u + B_b)\right) - K_N N_0 = 0 \qquad (7.15)$$

We will take the positive root and refer to it as $\hat{N}$. Then we have a two dimensional system,

$$\frac{dB_u}{dt} = B_0 F - \frac{\alpha_{max}}{K_\alpha + B_b}B_u + \frac{V}{A}\beta B_b - FB_u$$
$$+ (1 - \frac{B_b}{K_b + B_b})\frac{B_b}{B_u + B_b}(N_0 F - F\hat{N}), \qquad (7.16)$$

$$\frac{dB_b}{dt} = \frac{A}{V}\frac{\alpha_{max}}{(K_\alpha + B_b)}B_u - \beta B_b$$
$$+ \frac{A}{V}\frac{B_b}{K_b + B_b}\frac{B_b}{B_u + B_b}(N_0 F - F\hat{N}). \qquad (7.17)$$

7.3.3 Sensitivity Analysis: Correlation Coefficient, Spearman

It is difficult to interpret Pearson's correlation when the relationship between the parameter and the QoI is nonlinear. This is a difficulty

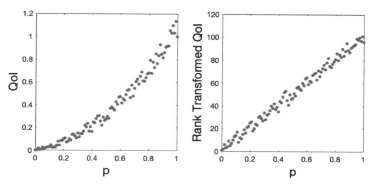

Figure 7.7: Example of raw data compared with rank transformed data.

that we have already seen when trying to use linear regression for scatter plots. In fact, it is exactly the same issue. The slope is not well interpreted for nonlinear relationships. One way to adjust this is to consider a rank transform where the actual values are replaced with ordinal rankings. This is sometimes referred to as a nonparametric method since the shape of the relationship is suppressed by the transformation. As long as the relationships are monotonic, rank transforming the data yields linear relationships. An example of this is shown in Figure 7.7 where the data are given on the left and the ranked transformed data are on the right. It is important to note that since we want this to be applicable in realistic settings, monotonicity may only be approximately satisfied and we will still have a reasonable approximation of linearity in the rank transformed data. We can then consider how close to linear (e.g. how strongly correlated are the relationships) by using the same calculation to determine Pearsons correlation, but on the ranked data. This gives a better estimate for nonlinear relationships than Pearsons [62, 33].

It is very important to note that monotonicity is essentially required in order to interpret any of the correlation coefficients properly. Just as linear regression to scatter-plots can be misleading if the scatter-plots are not approximately linear, rank transformation will not result in a relationship that is close to linear unless the untransformed data are monotonically related. Therefore, it is important to check this which can become difficult for many parameters.

For example, suppose we are interested in limiting the bacteria that attach to the wall of the chemostat over a fixed interval of time while maximizing the free bacteria produced. We need a measure that balances both of these. There are several ways to incorporate this into a QoI but one standard way is to consider the QoI,

$$\text{QoI} = \int_{t_0}^{t_1} \frac{B_u(t)}{B_b(t)} dt.$$

We would like to determine which parameters have the strongest affect on the QoI. Because the sign of the correlation coefficient indicates whether the QoI is positively or negatively correlated with the parameters. We have twelve parameters to rank – meaning we are sampling from a 12-dimensional space. We will use a hyper-rectangle centered at our nominal parameter set given in Table 7.1. How much variation should we assume? This is a complicated question that is often guided by some biological understanding. In the absence of any estimates for the variations in the parameters we are only guided by rules of thumb and some exploration. Typically one can choose the hypercube to have sides that are ± 5 % of the nominal value. So we would sample p_i in an interval $[.95p_i, 1.05p_i]$. Recall that we may not have any information about the probability distribution of the parameters, so we might choose to sample uniformly. We could write our own sampler, but there are many available packages that are very sophisticated and relatively simple to use. We will use the Python package from `pyDOE` (Python, design of experiments)[1] or the MATLAB command `lhsdesign` for uniform distributions and `lhsnorm` for normal distributions. The codes `Spearman.py` provide methods for ranking and sampling. `np.corrcoef` which is the built in numpy method. Similarly in MATLAB we can calculate this directly or use the command
`[RHO,PVAL] = corr(a',b','Type','pearson')`. In Figure 7.8 we show the Spearman ranking for the parameters of the Freter model and QoI$= \int_{t_0}^{t_1} \frac{B_u(t)}{B_b(t)} dt$.

---

[1]It is possible that you will have to install |pyDOE. Different platforms have different requirements explained on the webpage:
`https://pythonhosted.org/pyDOE/index.html`

Table 7.1: Nominal parameters for the Freter model.

Parameter	Value
N_0	$.1\ (\text{ml}^{-3})$
B_u^0	$.01\ (\text{ml}^{-3})$
K_N	$.1\ (\text{ml}^{-3})$
K_α	$.01\ (\text{ml}^{-3})$
K_b	$.5\ (\text{ml}^{-3})$
Y	.5 (dimensionless)
α_{max}	$.1\ (\text{t}^{-1})$
μ	$.1\ (\text{t}^{-1})$
β	$.01(\text{t}^{-1})$
F	$1\ (\text{t}^{-1})$
V	$1\ (\text{ml}^{-3})$
A	$1\ (\text{ml}^{-2})$

7.4   Cooperators, Cheaters, and Competitions

Bacteria are often used as simple biological models for complex phenomena. Often this means looking for examples that mimic ecological observations in other organisms. Bacteria and bacterial-specific viruses have been used to study disease processes where it is possible to have strict control on the bacterial phenotype, respiration, and interactions. Sometimes this is used to illustrate a broad principle which is the case in the principle of the tragedy of the commons. Here the thought experiment is a community with a common ground that is used for everyone to graze livestock. Each individual receives an increased benefit by grazing more livestock. Since there is no motivation to refrain from grazing, the common ground will be overgrazed leading to a collapse in the grazing for the community. So cooperation is required – where some sacrifice is given to see a benefit at the community level.

Microbes also cooperate in complex ways by producing substances that benefit the community by coordinating behavior as in biofilms. Also there are examples of products produced by individual bacteria

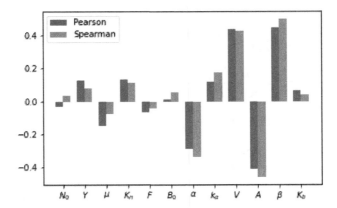

Figure 7.8: Comparison of Pearson and Spearman correlation for the Freter model.

that are used to help break down food sources increasing nutrient availability for the community at a cost for the individual. In fact, there are examples of populations that produce enzymes that are required for nutrient syntheses. What happens if a "cheater" type of bacteria enters the picture? Suppose a microbe is introduced that uses the product, leading to increased growth, but does not pay the penalty of production? Intuitively, this population will outcompete the non-cheater population and eventually eliminate them. This would cut-off the source of the enzyme and eliminate the nutrient, eventually dooming the population. The model described here is only one version of an interaction that can represent the tragedy of the commons but it does provide insight into how introducing individuals into populations can destabilize environments.

7.4.1 Model

A mathematical model of this situation requires a few pieces. The first is a food source, S, that needs and enzyme, E, to be processed into a form, P that can be consumed. The population of microbes consists of cooperators, B_1 that produce the enzyme and cheaters who do not, B_2. We will consider the system in a chemostat [58],

$$\frac{dS}{dt} = FS_0 - FS - \gamma SE, \tag{7.18}$$

$$\frac{dP}{dt} = \gamma SE - \frac{1}{Y_1}\frac{\mu_1 P}{K_1 + P}B_1 - \frac{1}{Y_2}\frac{\mu_2 P}{K_2 + P}B_2 - FP, \tag{7.19}$$

$$\frac{dE}{dt} = \alpha\frac{\mu_1 P}{K_1 + P}B_1 - FE, \tag{7.20}$$

$$\frac{dB_1}{dt} = (1-\alpha)\frac{\mu_1 P}{K_1 + P}B_1 - FB_1, \tag{7.21}$$

$$\frac{dB_2}{dt} = \frac{\mu_2 P}{K_2 + P}B_2 - FB_2. \tag{7.22}$$

The parameter α indicates the portion of nutrient consumption that goes directly to growth leaving the fraction, $1 - \alpha$, to go to enzyme production. Thus if α is close to zero, enzyme production does not penalize the cooperators very much. If α is close to one, most of the nutrient consumed by cooperators goes toward producing enzyme. Again, we need to specify initial conditions that are typically $(\bar{S}, 0, 0, \bar{B}_1, \bar{B}_2)$.

7.4.2 Analysis

Because we introduced this as a model of how cheaters can destabilize a population, we can look for this from the standpoint of stability. Notice that there is an equilibrium where there are no bacteria so that $B_1 = B_2 = 0$. These satisfy Equations 7.21 and 7.22.

At steady-state, Equation 7.20 implies $E = 0$ which is intuitive. If there are no bacteria to produce the enzyme it should eventually wash out of the system. Similarly, 7.19 implies that $P = 0$. Again, this makes sense. In the absence of enzyme there can be no processed nutrient. Without enzyme in the system, Equation 7.18 equilibrates to $S = S_0$, the source concentration. So there is definitely one steady-state $(S_0, 0, 0, 0, 0)$. Is it stable? We could explore the behavior numerically. Figure 7.9 shows several realizations for different parameters and initial conditions. We are guided by what we have learned from our simpler chemostat model. Namely, the growth rate plays a straightforward role in whether bacteria are maintained in the chemostat or whether they washout.

We show four realizations of the dynamics. First, with no cheaters, the system can evolve to a nontrivial steady-state for the cooperators.

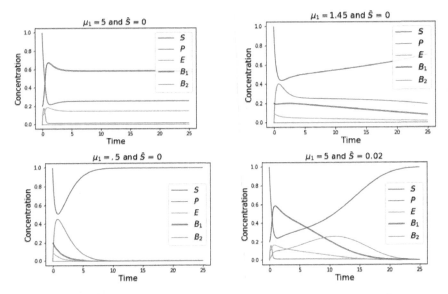

Figure 7.9: Comparison of the dynamics for the cheater/cooperator system in a chemostat. The base parameters are: $F = 1$, $S_0 = 1$, $\gamma = 20$, $Y_1 = Y_2 = 1$, $\mu_2 = 5$, $K_1 = K_2 = .05$, $\alpha = .2$. We change the initial conditions between the cheater case ($\bar{S} = .02$) and the case with no cheaters ($\bar{S} = 0$). We also allow $\mu_1 = 5, 1.45, .1$.

As the growth rate of the cooperators decreases, this steady-state is lost and the wash-out solution is stable. Without cheaters, this is a slightly more complicated model than the original chemostat model, but the dynamics are quite similar.

When cheaters are added, the system is destabilized and only the washout case exists. Notice that it appears that the system tends to the washout steady-state if *any* cheaters are introduced. We should ask whether this is true in general or something that we just happened to see for a specific parameter set.

7.4.3 Sensitivity Analysis: Sensitivity in Time and Partial Rank Correlation Coefficient (PRCC)

In this model, if there are any cheaters in the system the washout steady-state is the only stable state so the system always evolves to the washout state. Therefore understanding the a QoI that only depends on the steady-state does not require the differential equations. This is an important observation since there are many examples of sensitivity analysis that focus only on the asymptotic (long time) behavior. In this example we will explore sensitivity in time. Suppose our goal is to maintain the cooperator population as long as possible. That is we want to use $QoI = B_1(t)$. We know that as time goes on $B_1(t) \to 0$, which means the variance $B_1(t) \to 0$ as well, so we expect our measures all to be small.

One option would be to analyze the eigenvalues of the linearize system around the steady-state. The eigenvalues must all be negative since the steady-state is stable. So we can ask how the maximum (the least negative) of these eigenvalues depends on the parameters. This helps explain how the stability depends on parameters but does not describe how the evolution of the system *toward* the steady-state depends on the parameters. For that we will use sensitivity in time which is relatively self-explanatory. We repeatedly perform sensitivity analysis at discrete times, $(t_1, t_2, ...t_n)$. We then look at how the values change in time. For our measure, we will introduce a new method that is one of the most used measures of sensitivity.

Recall that Spearman's correlation coefficient measured the scaled correlation between a parameter and the QoI. This is a reliable measure for problems where the parameter, QoI relationship is close to linear. Pearsons correlation coefficient uses rank transformation to turn nonlinear relations into linear, where the correlation coefficient again quantifies the relationship. But this also includes variations in the QoI due to variations in other parameters. Partial correlation coefficients discount this to isolate the portions of that only describe the single parameter/QoI relationship. This is often attributed to Kendall [37, 24, 41].

From a geometric point of view, the idea is to restrict the QoI/-parameter relationships to one plane at a time. The generalization of scatter plots to multi-dimensions is sometimes referred to as a point cloud. Looking at the projection of the point cloud onto specific planes produces scatter plots that we have seen before (see Figure 7.10). This projection + regression discounts correlations that are hidden through parameter combinations.

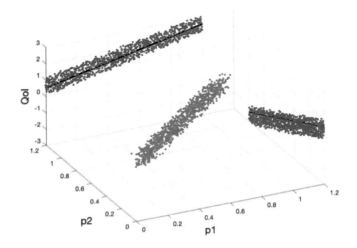

Figure 7.10: Point cloud of a QoI that depends on two parameters along with projections on the the (p_1, QoI) and (p_2, QoI) planes.

Combining ranked correlation and partial correlation provides a very robust sensitivity method referred to as Partial Rank Correlation Coefficient. In Figure 7.10, we should scatter plots of a QoI that depend on two parameters, p_1 and p_2. We also show the projections of this data onto the coordinate planes (p_1, QoI) and (p_2, QoI). The linear regression of the projected data is also shown. The relative the the variances of these lines are the partial correlation coefficients. By projecting the data onto the (p_i, y)-planes, we discount the contribution of the other parameters on the correlation. We can then estimate the correlation in the restricted plane.

There is a formula similar to that in Equation 7.23, where the variation in the QoI due to p_j is discounted when calculating the correlation between p_i and the QoI. Recall that the Spearman correlation between X and Y is

$$r_{X,Y} = \frac{\sum_{i=1}^{n} (X_i - \bar{X})(Y_i - \bar{Y})}{\sqrt{\sum_{i=1}^{n} (X_i - \bar{X})^2 \sum_{i=1}^{n} (Y_i - \bar{Y})^2}}.$$

The partial rank correlation coefficient for a single parameter, p_1 and for a QoI that depends on parameters (p_1, p_2) where the parameters have been rank-transformed is,

$$prcc_{p_1} = \frac{r_{p_1,QoI} - r_{p_1,p_2} r_{p_2,QoI}}{\sqrt{(1 - r^2_{p_1,p_2})(1 - r^2 p_2, QoI)}}. \tag{7.23}$$

In general, to discount all the interactions we subtract the sum of all covariances between p_1 and p_j. There are many ways to implement this but we will use the specific Python and MATLAB implementations in the codes. Python implementation is based on the `pingouin` that is a widely uses statistical package[2]. There are many other implementations that can be found, however. MATLAB has a "native" command for partial correlations.

We will recall the workflow that is the same for all of the methods in the chapter:

- Sample parameter space:
  This can be done using Latin-Hypercube sampling, Monte Carlo or other methods. There is not a definite number of samples to draw, but typically for models that are reasonably computationally simple (e.g. not large or mutlidimensional), all of the methods are relatively efficient.

- Evaluate the model at all parameters:
  You will need to manipulate these values so it is typical to store them in an array.

- Evaluate the correlation:
  whether that is Spearman, Pearson, PRCC, or other statistical methods may require rank transformation.

If we want to examine sensitivity in time, we perform this at specific times. For example in Figure 7.11, we define the $QoI = B_2(t_{final})$ and rank the parameters for several different values of t_{final}.

Sensitivity in time provides information about the time-scale of the biological processes that dominate the specific behavior. This can be very important when working with experimentalists since this helps

---

[2]The website `https://pingouin-stats.org/index.htm` — has information about installation and implementation.

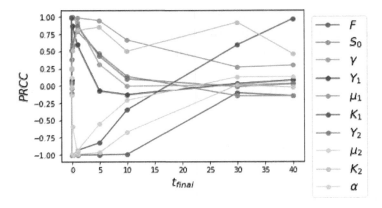

Figure 7.11: PRCC values in time for the cheater/cooperator model. We see that there is an abrupt re-ordering within the first time unit. They there is another transition around $t = 10$. Several parameters that were sensitive initially are not sensitive at later times (e.g. F).

define time-scales for the experiments. For example, if there is a rapidly changing sensitivity at the beginning of an experiment, taking observations too far apart may miss crucial information. Additionally, if the aim is to control the behavior, the most sensitive parameters provide the most efficient tools in theory. If the sensitivity rankings change in time, the best control targets may change in time as well.

7.5 State of the Art and Caveats

PRCC is a widely used ranking method. It is quite robust and simple to implement. One advantage that PRCC has over other global methods is an interpretation of the sign of the correlation for nonlinear models. The intuition is that PRCC values for a parameter that are close to one imply that a small increase in that parameter leads to a relatively larger increase in the QoI. Similarly, if the PRCC value for a parameter is close to -1, a small increase in the parameter leads to a relatively larger decrease in the QoI. This can be quite informative when trying to control model outputs.

The two caveats that are important to reiterate are that for correlation coefficients to be robust, the relationships need to be close to linear. For PRCC this means that the parameter/QoI relationship needs to be monotonic so that the rank transformation leads to a linear relationship. This often requires visual inspection of scatterplots and reduction of the parameter space, similar to what was done for scatterplots and regression.

Second, there is essentially no theory regarding how many samples are required. There are estimates and rules of thumb (which we have not included precisely because there is not a solid theoretical justification). However, PRCC requires far fewer samples that many other methods. It is good practice to vary the number of samples before concluding that the ranking values are acceptable.

7.6   Problems

Problems 7.1 *Consider the function* $f(c) = \frac{\mu c^n}{k + c^n}$.

(a) Sketch the graph of f for varying values of n.

(b) For $n = 1$, what is $f(k)$? Show this on the graph.

(c) Use Taylor's theorem to write the linear approximate of $f(c)$ near $c = 0$ so that $L(c) \approx f(c)$.

(d) Use numerical methods to compare the solution of the differential equations

$$\frac{dx}{dt} = f(c)x,$$

and

$$\frac{dy}{dt} = L(c)x,$$

for different values of c and n. Is the agreement better for large of small c? Why?

Problems 7.2 *Consider the matrix of parameters for the case of two parameters:*

$$\begin{bmatrix} p_{1,1} & p_{1,2} \\ p_{2,1} & p_{2,2} \end{bmatrix}$$

With QoIs

$$\begin{bmatrix} y_1 \\ y_2 \end{bmatrix}$$

(a) *Expand the equation for Pearson's correlation (Equation 7.6) for the sensitivity of y_1 with respect to p_1.*

(b) *Show that this is always between -1 and 1.*

(c) *Suppose we have the numerical values for the parameters and outputs,*

$$\begin{bmatrix} .1 & 3 \\ .4 & 1 \end{bmatrix}$$

With QoIs

$$\begin{bmatrix} 3 \\ 1 \end{bmatrix}$$

Compare Pearson's and Spearman's Correlation – this can be done by hand or using the computer codes. How do these compare with PRCC? Why are they the same?

Problems 7.3 (a) *Suppose the source of nutrient for the Freter model is periodic in time, so that the source of nutrient is $N_0 = N_0 \sin(\lambda t)$. Show solution curves for varying λ.*

(b) *How does this change the Spearman ranking where the QoI ratio of the total of the two bacterial concentration?*

(c) *One of the most overly-simplified parts of this model is the detachment which is assumed to be constant. It is more realistic for β to depend on the flow rate and the density of the attached bacteria since these are typically in a biofilm and much harder to remove. Change the detachment so that it is a decreasing function of B_b. How does this change the sensitivity ranking where the QoI is the maximum bound bacteria in a given length of time?*

Problems 7.4 *We did not demonstrate that the parameter/QoI relationship was linear for PRCC in time applied to the cheater/cooperator problem. Adjust the codes to display the scatter plots and discuss the implications. Are there parameters where using a linear correlation is not appropriate? If so, what happens if you reduce the range of those parameters to restrict to regions where the relationship is monotonic?*

Problems 7.5 *(a) Determine the stability of the two steady-states of the chemostat equations.*

(b) *Explore the effects of adding a periodic source to the nutrient by altering the chemostat numerical script.*

(c) *Does the period change the behavior? Can this be quantified? Add the period as a parameter and compare the ranking of this parameter with the other parameters using PRCC.*

Problems 7.6 *Consider how to change the chemostat and tragedy equations to vary F. Define a new flow rate, $F = F\cos(\lambda t)$. This introduces a new parameter λ.*

(a) *Implement this F for both the chemostat and tragedy codes. Show the behavior for different values of λ.*

(b) *How does λ rank when the QoI is the maximum of the bacterial concentration in the chemostat? How do Pearson, Spearman, and PRCC compare?*

Problems 7.7 *(a) Use any method to determine the stability of the washout steady-state of Equations 7.18–7.22.*

(b) *Suppose that cheaters revert to cooperators at a fixed rate, κ. Change the equations and determine if the washout steady-state is stable for all values of κ.*

(c) *Define a QoI that quantifies the time that B_2 is maximum. Rank the parameters with respect to this QoI. Be sure to interpret this biologically.*

8

Circulation and Cardiac Physiology

Cardiac models are widely used to diagnose heart defects. This is crucial for drug development and treatments. We describe models of blood circulation and cardiac physiology. Sensitivity methods topics include Morris Screening and parameter reduction.

8.1 Historical Background

Mathematical models of cardiovasculature dynamics are among the most diverse models in biology. This has much to do with the complexity of the combination of physiology, physics, and chemistry in an environment where direct observation is difficult. The heart requires coordination between the chemistry that drives an electrical signal and the induced muscle contraction. There are also numerous passive physiological structures such as valves that help control the blood flow by interacting with the flowing blood by opening and closing pathways. As the electrical signal is transformed into mechanical contraction, pumping blood throughout the body the blood transports material that controls the electrical signal. Blood circulation is responsible for transporting nutrients and waste products (oxygen/carbon dioxide exchange), circulating signals (hormones), managing the transport of portions of the immune system (leukocytes), managing cellular damage (wound repair and clotting), and regulating temperature.

Understanding the role of the blood circulation was intimately tied to the transformation from the ancient understanding of physiology represented by Galen of Pergamon through William Harvey and into the modern era the mid-1800's. Galen lived around 200 AD and was aware of the cardiopulmanary system viewed it as a one-way transport system that brought food processed in the liver through the body as it was transferred into blood which was, in turn, transferred into flesh. This framework organized his observations of surgery as well as dissection where he noted structural differences between arteries and veins, the connection between respiration and blood transport, and the motion of the heart in relation to the blood.

The historical basis of physiology often cites William Harvey as the key character in the development of the modern era. In the late 1500's to early 1600's, Harvey was both a working physician – the personal physician of James I – and an active researcher. Harvey was able to observe the motion of the heart in living animals and note that there was more blood being processed in the hard than could be absorbed by the tissue. He was demonstrated the one-directional motion of the blood and formulated the modern picture of the heart as a pump that drives the circulation of blood throughout the body. Shortly afterward, in 1692, Max Perutz discovered the structure of hemoglobin which carries oxygen in the blood.

It is important to note that between Galen and Harvey there were many other scientists who also studied the connection between blood flow and oxygen transport. In the 1200's the Syrian physician Ibn al-Nafis noted that "Pulmonary arterial blood passes through invisible pores in the lung, where it mingles with air to form the vital spirit and then passes through the pulmonary vein to reach the left chamber of the heart" – which is essentially a correct view of circulation theory. His observations were essentially unknown in the west so it was another 400 years before Harvey was able to accumulate enough evidence and support to move beyond the works of Galen.

So where does this leave mathematical modeling? Mathematical models have been used at multiple levels from the single cell up to geometrically accurate representations of the muscle contractions that drive the blood circulation. Electrochemical models of cardiac dynamics are among the most intricate models in biology. Even neglecting the spatial aspect of the models – which vastly complicates the blood flow and coupling– because there are thousands of channels in each cell within the heart, each of which alters the dynamics of ion transfer and plays a role in regulating the heart beat, models can become very large – as in over 100 differential equations where details about each ion channel is resolved. It is not entirely clear what one should do with these types of models since understanding the roles of the parameters in the predictions is very difficult to understand.

Most of the sensitivity methods that we have explored rely on some version of sampling. Even those that do not depend on the nominal point in parameter space that we are using as our estimates. So what happens to our sampling as we add more parameters (e.g. extend the parameter space by an additional dimension? This difficulty is referred to as the *Curse of Large Dimensions*. There are two views that are important. The first is percent coverage obtained by sampling. If we have 100 samples in a one dimensional interval $[0, 1]$, each sample covers 1% of the parameter space. If we add a parameter so that our parameter space is two dimensional and keep 100 samples each sample covers only .01% of parameter space. Add another dimension and each sample covers only .001% and so on. Therefore the number of samples increases exponentially as the dimension increases.

Secondly, it turns out to be very difficult to sample efficiently and still reach the corners of the parameter space. This is also relatively simple to see. Consider a one dimensional parameter space – that is an interval $(0, 1)$. The entire interval is within a one dimensional "ball" of radius 1. Here we are using the term "ball" in the technical sense where an n-dimensional ball of unit radius is the set of all points whose distance from the center is less than or equal to 1. If we consider the ratio of the length of a unit ball in one-dimension to a unit interval, the ratio is 1. Now consider the ratio of the area of a two-dimension unit ball (e.g. a circle) to the area in a two-dimensional hypercube (e.g. a square). This ratio is $\frac{\pi}{2^2} \approx .78$. If we continue, the ratio of the unit ball in three-dimensions (e.g. a sphere) to the unit hypercube (e.g. a cube) we have $\frac{\frac{4}{3}\pi}{2^3} \approx .52$. This means that large dimensional hypercubes are almost entirely corners – that is hyper-volumes that lie outside of the unit sphere. A visual of this is shown in Figure 8.1. Sampling in these regions becomes more and more difficult as the dimension increases. This means that standard random sampling can miss large portions of your parameter space – and these are exactly the regions where it is likely that the true parameters lie since these are the largest portions of parameter space.

There are two concepts that we will explore here in connection to the curse of large dimension. The first concept relates to sampling – are there methods that are more suitable for sampling in large dimensions? How do these methods compare? We have only used two different sampling methods: "random" and Latin Hypercube Sampling. In the context of large dimensions, where the main issue revolves around ensuring full coverage of parameter space, random sampling is very problematic since it has no way to ensure that corners of parameter space are reached. Suppose we want to sample and ensure that the distance between our samples and a particular point in parameter space are all about 1% of the length of one of the sides of the hypercube

Richard Bellman (1920–1984)

Richard Bellman was an applied mathematician who spent time in academia at Princeton and Stanford. In 1952 Bellman moved to the RAND corporation until 1965 when he moved to University of Southern California. Bellman wrote more than 600 manuscripts and 40 books. He wrote more than 10 papers per year for the last 10 years of his life. He pioneered dynamic programming which is an optimization technique that breaks down problems into smaller subproblems. These problems are solved recursively so they solution can be refined. It was in this context that he discussed the exponential increase in volume as the dimension increased.

An example of the idea of dynamic programming can be seen in matrix product ordering. We know that the product of three matrices can be calculated as $XYZ = X(YZ) = (XY)Z$. But is there a difference in the efficiency depending on the order that is taken? It turns out that there is. If the X, Y, and Z are 10×5, 5×10, and 10×5 matrices. Then the result is an 10×5 matrix. It takes $5 \times 10 \times 5$ steps to calculate YZ which is a 5×5 matrix. Multiplying this by X takes $10 \times 5 \times 5$ steps. Doing it the other order requires $10 \times 5 \times 10$ steps to calculate XY, which is a 10×10 matrix. Multiply this by Z requires $10 \times 10 \times 5$ additional steps. So the first way takes $250 + 250$ while the second way takes $500 + 500$ steps. This can be generalized recursively to find the optimal multiplication of n matrices.

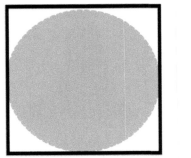 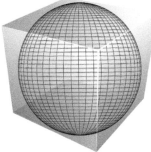

Figure 8.1: Unit ball compared to unit cube in one and two dimensions. The relative volume between the sphere and cube increases as the dimension increases.

containing our parameters – we can assume that the cube has sides of unit length. In one-dimension we need 100 samples. In two-dimension we need 100^2 samples. In n-dimensions we need 100^n samples. This rapidly becomes intractable. In contrast, Latin Hypercube Sampling is proportional to n where the constant of proportionality is related to the refinement in each direction. While neither of these arguments is a proof, they are compelling enough to argue that Latin Hypercube Sampling is better are spreading out our samples.

One other concern that has been touched on but not addressed yet is convergence. What happens to our sensitivity estimates as the number of samples increases? Certainly at some point we should have all the information that is needed and the estimates should stop changing. We have not focused on this since many of our methods have been visual (scatter plots) although the rule of thumb for PRCC is to use at least $\frac{4}{3}n$ where n is the number of parameters. This is only an estimate and one should be cautious about trying to minimize the number of samples used. For most of the models that we are focusing on this is not the main concern; however, in practice this may become a limiting factor.

Trying to limit the number of samples needed leads to the concept of "freezing" parameters. This is actually one of the more historical aspects of sensitivity. If we were able to identify the *insensitive* parameters, we would identify parameters that do not have an impact on our predictions. These parameters could be fixed in our analysis since they

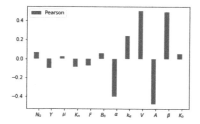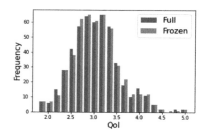

Figure 8.2: The left plot shows estimates of the sensitivity indices –
specifically for the Freter model in Chapter 7. The right plot shows a
comparison between the histograms of the QoI varying all parameters
and varying only the five parameters with the largest sensitivity.

do not alter the outcome. This has the effect of reducing the parameter
space needed to capture the behavior.

There are many ways to verify that freezing the parameters does
not alter the predictions. First of all it is important to understand that
which parameters are insensitive is tied to the QoI – changing the QoI
requires recalculating the sensitivity. Second, one common method for
comparing the model behavior while all parameters are varied to the
behavior for the reduced parameter space is to compare the output
statistics. In Figure 8.2, we show a comparison between histograms
for the output QoI while varying all parameters and for varying only
the sensitive parameters (i.e. full model versus frozen model).

There are methods for quantifying the differences in histograms
such as the area between the two PDF's, but it is clear that most of the
information is captured in the sensitive parameters. We have focused
on the most sensitive parameters since these are ones that need to be
estimated carefully and are potential targets for controlling the system.
We can also conclude that we do not need to vary the insensitive parameters.

We will briefly mention one other method for generating samples
that is often used. Because random sampling can easily miss regions
of parameter space it is typically avoided. Another way to look sampling methods that miss portions of parameter space is to introduce the
concept of discrepancy which is a measure of the density of points in
space. High discrepancy means there are a lot of sparse points while

low discrepancy implies the density is relatively the same through-out space. Random sequences are high discrepancy. But what about a grid-based uniform sample? Unless the refinement of the grid is high enough the discrepancy of these samples if often too high. To address this there are numerous low discrepancy sequences that can be used. We do not go into their constructions here but the terminology appears in many numerical codes – several well studied examples are Sobol', Halton, and van der Corput sequences.

8.2   Blood Circulation Models

Because of the complex regulation, understanding the circulatory sys-tem can become very involved. There are numerous examples where fluid dynamic models have been used to describe the interaction be-tween the beating heart and the resulting flow of blood. These models include the contractile mechanisms of the heart muscle that drives the fluid as well as the force the fluid exerts on the heart and blood vessels. But what if we are interested in a larger scale question regarding the relationship between the flow of blood and the pressure exerted?

Beginning in the 1950's compartmental models were introduced to lump different parts of the circulatory system and use the analogy with circuits to develop dynamic models. These models have grown to include multiple compartments including multiple heart chambers, ar-terial and venous blood vessels, and the peripheral blood supply. Here we will make several simplifications that make the model tractable. The main simplifications are to neglect time variations – this means we are only considering averaged circulation. We will also use a two com-partment model – one that includes the heart and the systemic blood supply (see Figure 8.3).

Following [36], we will simplify the dynamic model first intro-duced in [23] and develop an algebraic model for the steady-state re-lationship between blood pressure and the volume of blood moved by the heart.

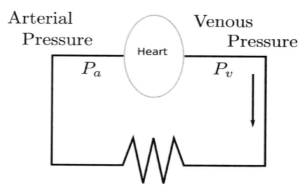

Figure 8.3: Schematic of a two compartment model of the circulatory system. The heart is a compliant vessel that pumps blood through the venous system with resistance, R.

8.2.1 Model: Algebraic

The flux of blood through a blood vessel, Q, is related to the pressure drop, ΔP that drives the blood. The simplest model is analogous to the current through a resistive circuit using Ohm's Law: Current $= \frac{\text{Voltage Drop}}{\text{Resistance}}$. By analogy,

$$Q = \frac{\Delta P}{R} = \frac{P_0 - P_1}{R}. \tag{8.1}$$

The resistance of flow through a cylindrical vessel is inversely proportional to the square of the cross-sectional area so that halving the radius increases the resistance to flow by a factor of 8. However, this calculation is based on the assumption that the radius is constant. Blood vessels are compliant and can expand and contract depending on the blood flow. The simplest model of this relates the cross-sectional area of the blood vessel to the pressure, $A = A_0 + cP$. If we go back to the flux/pressure relationship for a compliant vessel we find a non-Ohmic relationship. This is a more complicated relationship than in Equation 8.1,

$$Q = \frac{(P_0 - P_1)}{R}\left(1 + \gamma(P_0 + P_1) + \frac{\gamma^2}{3}\left(P_0^2 + P_0 P_1 + P_1^2\right)\right). \tag{8.2}$$

Here R is inversely proportional to A^2 and the compliance, γ, is inversely proportional to A. Rather than writing the pressure drop as ΔP

we are writing this as the difference between the input pressure, P_0 and the output pressure, P_1. If γ – that is with no compliance, we recover the original version.

A similar calculation relates the total volume in the vessel and the pressure drop between P_0 and P_1,

$$V = V_0 \left(1 + \frac{\gamma}{2}(P_0 + P_1) + \frac{\gamma^2}{6}(P_0 - P_1)^2 + \gamma^3 \xi(P_0, P_1) \right), \quad (8.3)$$

where $\xi(P_0, P_1)$ is a function of terms involving cubic powers of P_0 and P_1.

Restating the flux and volume in terms of the compartmental model in Figure 8.3, the flux through the compliant vessel compartment is,

$$Q_{vessel} = \frac{(P_a - P_v)}{R} (1 + \gamma(P_a + P_v) \dots$$
$$+ \frac{\gamma^2}{3} (P_a^2 + P_a P_v + P_v^2)). \quad (8.4)$$

while the volume of blood within the vessel is,

$$V_{vessel} = V_0(1 + \frac{\gamma}{2}(P_a + P_v) + \frac{\gamma^2}{6}(P_a - P_v)^2). \quad (8.5)$$

If we model the heart as a compliant vessel and break down the behavior during a single beat to include systolic and diastolic dynamics the total cardiac output for a single beat is,

$$Q_{heart} = F(V_{max} - V_{min} + C_d P_v - C_s P_a), \quad (8.6)$$

where V_{max} and V_{min} denote the maximin and minimum heart volume. The compliances of the heart depends on which pressure dominates (i.e. where in the beat we are looking) and are denoted C_d and C_s. Finally, F represents the number of beats in the measured time.

Since the flux in the heart must equal the flux in the blood vessel (that is blood is incompressible), $Q_{vessel} = Q_{heart} = Q$. Similarly, the total volume of blood is constant so we can view V_{vessel}, V_{max}, and V_{min} as constants. Therefore Equations 8.4, 8.5, and 8.6 define the unknowns Q, P_a, and P_v in terms of the parameters R, γ, V_{vessel}, V_0, F, V_{max}, and V_{min}.

8.2.2 Analysis

These are algebraic equations can be solved numerically. We can notice a few things about this model. First, if neither the vessel or the heart compartments are non-compliant, the volume of blood in the vessel is constant. We find that $Q = F(V_{max} - V_{min})$ which implies that the pressure drop in the vessel must balance this, $P_a - P_v = RQ$. In the non-compliance case, $V = V_0$ and we cannot determine P_a and P_v uniquely.

What if we assume that γ is very small? Then $\gamma^2 << \gamma$ and we have an approximation that is valid for a range of γ's near zero. This becomes,

$$Q = \frac{(P_a - P_v)}{R}(1 + \gamma(P_a + P_v)), \tag{8.7}$$

$$V_{vessel} = V_0(1 + \frac{\gamma}{2}(P_a + P_v)), \tag{8.8}$$

$$Q = F(V_{max} - V_{min} + C_d P_v - C_s P_a). \tag{8.9}$$

We can show that this has a unique solution. This gives is hope that the nonlinear model has a solution as well. Additionally, solving the linear system provides a very good initial guess that we can use in our numerical solver. To solve this numerically, we can revisit codes that we built from before when we were looking for steady-states. In the next section, we will consider how variation in these parameters change the flux and blood pressure.

8.2.3 Sensitivity Analysis: Sampling Methods

As we discussed previously, there are many ways to sample. Typically, high blood pressure refers to the arterial blood pressure. In part, this is because arterial pressure is substantially higher than venous blood pressure in healthy individuals. We will use P_a as the QoI. Rather than introduce a new quantification here we will look at differences between the histograms (and estimated PDF) as we vary the sampling method.

We can choose between several sampling methods. For simplicity, we can assume that the parameters are distributed uniformly. We then pull samples from the distribution in several ways. We could use Monte Carlo sampling which is a version of randomized sampling, LHS sampling, uniform sampling, or low discrepancy sampling. With any of

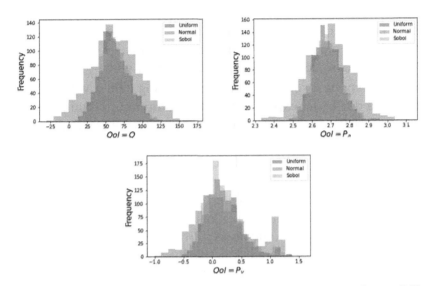

Figure 8.4: Histograms of the different QoI's across three different sampling methods.

these methods, we can select a fixed number of samples, determine the QoI (P_a) and arranged these in a vector. We then plot the histogram of the QoI's which can be used to approximate the PDFs. In Figure 8.4 we compare the outputs for Q, P_a, and P_v using three different sampling methods and the fixed point codes. We see that broadly there is agreement; however, there are some clear differences that might need to be considered.

8.3   Cardiac Physiology

The electrical signals that control the contraction and relaxation of a cardiac cell are driven by ion exchange in a manner very similar to the regulation of signals along axons. In response to changes in the voltage across the membrane of a cell, ion specific channels open and close allowing ions to move into and out of the cell. We will not discuss the details in this chapter (instead see Chapter 9). Instead we will note that changing ions are a current, which is the rate of change of voltage. So there is a direct link between cardiac models – or more generally models of excitable media – and electrical circuits.

Models of cardiac dynamics typically begin with $\frac{dV}{dt} = \frac{\sum I_{ion} + f(t)}{C_m}$, where $f(t)$ may be a source from a neighboring cell, a voltage potential imposed during an experiment or the addition of certain ions. The capacitance of the membrane is denoted C_m. The currents depend on the voltage and the state of the ion-specific channels. Thus the voltage equations are augmented with equations describing the state of the channels, or "gating variables". These channels regulate the flow of calcium, chlorine, sodium, potassium, and multiple others.

The time-scale of opening and closing of the channels is one of the main characteristic of the ionic flows. Several of the species equilibrate quickly and others do not. Traditionally, this allows for approximations to analysis the behavior on different time-scales. For our purposes, this implies that the differential equations may need to be handled with some care. Rapid changes in variables can cause difficulties in numerical approximations of solutions, so if we move too far in parameter space we may expect to see issues that are purely numerical.

8.3.1   Model: Noble

Models if the electrical signals have become more detailed as more research has specified the behavior of different ionic channels. Additional models may include internal structures such as the endoplasmic reticulum that store and release ions such as calcium, also increasing the complexity. There are examples of models that include more than 50 variables and 100's of parameters. These models are highly refined, connected with specific experiments, and beyond the present presentation. Instead, we will describe one of the progenitor models – one that set the stage for decades of refinement. This model 1962 model, often referred to as the Noble model [51], focuses on specific types of cardiac cells: Purkinje fibers.

These fibers make up the bulk of the conductive system within the heart. Along with the SA node, AV node and specific bundle cells, Purkinje cells coordinate and transmit the electrical signal that regulates the contraction of the cardiac smooth muscle cells. Typical electrical signals measured using EKG's are action potentials that show how the voltage changes in time, so the goal is to describe the evolution of the voltage action potential.

The equations govern the voltage, v, that depends on ionic concentrations of sodium, I_{Na}, potassium, I_k, and an aggregate current called the leak or background current, I_{bk}. The background current is a catch-all that includes ions such as chloride.

The currents depend on the state of the activation and inactivation gates. The gating variables alter the currents and are defined by differential equations. One can think of the gating variables as the proportion of the channels that are open. This proportion changes depending on the voltage. The time constant associated with the dependence specifies the gating variables for each ionic species. A general equation for a gating variable, G can be written,

$$\frac{dG}{dt} = \alpha_G(1 - G) - \beta_G G. \tag{8.10}$$

Notice that if $\alpha_G = 0$, $G \to 0$ while if $\beta_B = 0$, $G \to 1$. The α_G and β_G coefficients depend on V and are often referred to as time-constants because they set the time-scale of opening and closing–even though they are implicitly functions of time.

We can collect the equations for the voltage, the three currents, and time-constants. The voltage equation is,

$$\frac{dv}{dt} = -\frac{(I_{Na} + I_K + I_{bk})}{C_m}. \tag{8.11}$$

The currents include the sodium current, I_{Na}, the potassium current, I_K and the background or leak current, i_{bk}. The sodium current is,

$$I_{Na} = (400000m^3h + 140)(v - E_{Na}), \tag{8.12}$$

where there are two gating variables, m and h. These follow the dynamics,

$$\frac{dm}{dt} = \alpha_m(1 - m) - \beta_m m, \tag{8.13}$$

with time constants regulating opening,

$$\alpha_m = \frac{100(-v-48)}{e^{\frac{(-v-48)}{15}} - 1}, \tag{8.14}$$

and closing,

$$\beta_m = \frac{120(v+8)}{e^{\frac{(v+8)}{5}} - 1}. \tag{8.15}$$

The h gate is similar,

$$\frac{dh}{dt} = \alpha_h(1-h) - \beta_h h, \tag{8.16}$$

$$\alpha_h = 170e^{\frac{(-v-90)}{20}}, \tag{8.17}$$

$$\beta_h = \frac{1000}{1 + e^{\frac{(-v-42)}{10}}}. \tag{8.18}$$

The potassium current is given by,

$$i_K = \left(1200e^{\frac{(-v-90)}{50}} + 15e^{\frac{(v+90)}{60}}\right)(v - E_k) \tag{8.19}$$

$$+ 1200n^4(v - E_k), \tag{8.20}$$

with gating variable, opening and closing time constants,

$$\frac{dn}{dt} = \alpha_n(1-n) - \beta_n n, \tag{8.21}$$

$$\alpha_n = \frac{0.1(-v-50)}{e^{\frac{(-v-50)}{10}} - 1}, \tag{8.22}$$

$$\beta_n = \frac{e^{(-v-42)}}{10}. \tag{8.23}$$

The background anion current is instantaneously adjusted based on the voltage,

$$i_{bk} = 75(v - E_{an}). \tag{8.24}$$

Clearly many of the constants have been fixed – these are based on empirical estimates used to match EKG readouts. Therefore we should consider values such as 400,000 and 140 in Equation 8.12 as nominal parameter values and therefore only estimates. We can legitimately question how the predictions depend on these values as well as values for other parameters (e.g. E_{an}).

It is important to note that these equations were developed prior to a full understanding of the role of both sodium and calcium – the latter in particular plays a fundamental role in the contractile process. However, even with these difficulties, the Noble model was able to capture much of the dynamics. For example, the Noble model did not require a pacemaker to drive the oscillations. Even more importantly, the Noble model set the stage for successively more detailed and accurate models []. These models where crucial for the development of drugs aimed at regulating the heart beat and treating acute and chronic conditions that can lead to heart attacks, arrhythmias, fibrillation, and other significant heart conditions.

8.3.2 Analysis

One can understand the model in several ways. One general way is to use fast/slow analysis or quasi-steady-state analysis as we did in 7 for the Freter model. This method requires determining which variable changes slowly relative to the other variables. We then approximate the system assuming the rate of change of this is zero (i.e. the left-hand-side of the equations). This leads to an algebraic relationship that can reduce the dimension of the problem.

For example, we could assume that n equilibrates quickly so. that $\alpha_n(1 - n) - \beta_n n \approx 0$. Then $n = \frac{\alpha_n}{\alpha_n + \beta_n}$. We then have a system of two differential equations (V and h and the algebraic equation for i_K. We can then use phase-plane analysis to broadly understand the dynamics.

We can also do direct simulations. What we see from this is that this model exhibits periodic behavior that is self-sustaining (see Figure 8.5). This pattern is called an action potential. This is a key feature of the Noble model. It turns out that it is incorrect in the sense that mechanism that drives the periodicity in the physiological setting is not faithfully recreated in the model. However, for the purposes of demonstration, we will use this model and consider features that are

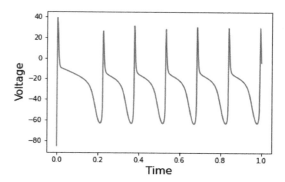

Figure 8.5: An example voltage trace from the Noble model.

important that we can highlight to consider the impact of parameter variations.

An action potential is a rapid rise and subsequent fall in the membrane potential. Action potentials have broad characteristics that indicate the excitability of the tissue. There are several stages of a generic action potential including the rapid polarization/depolarization followed by a refractory period and return to rest. For some physiological settings this is highly refined. For example, the action potential associated with the beating heart is directly related to the EEG signal (see Figure 8.6).

There are several key features that are connected to clinically important diagnosis. One of the important features of an electrocardiogram (that measures V as a function of time) is the regularity of the voltage trace. Irregularities in the action potential are referred to as arrhythmias and can lead to heart attacks. We could quantify this many different ways, we could count the number of peaks in a period of time. However, this can cause difficulties because the peak of the action potential is not fixed for different parameters. It is actually quite complicated to consider varying periodic behavior and leads to methods that are far beyond the scope of this book. We will use a relatively simple way to estimate the frequency. The Fourier transform moves a function of time to the "frequency" domain – the independent variable of an Fast Fourier Transform is a frequency. We take the Fourier transform of our voltage trace, determine the maximally occurring frequency and

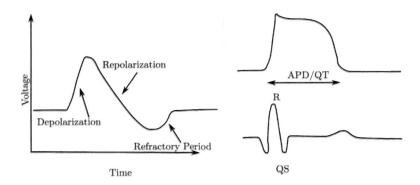

Figure 8.6: A schematic of a generic action potential. There are specific features of a cardiac action potential that are important such as the action potential duration (APD). The action potential for a cardiac cell can be connected with measurements taken using an electrocardiogram. The PQR complexes of an EEG help diagnose defects and lead to treatments.

use this as a measure of the frequency of the voltage oscillation. An example work-flow is show in Figure 8.7.

8.3.3 Sensitivity Analysis: Morris Screening

We will now show an example of using sensitivity analysis to reduce the parameter space. To do this, we need some form of sensitivity

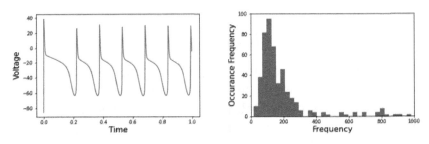

Figure 8.7: The trace of the voltage curve along with the Fourier transform of the signal. The peak of the Fourier transform is approximately 100.

ranking – we will introduce a screening method called Morris screening [34, 48]. Screening methods are typically less computationally intensive than other methods and the goal is to separate out sensitive and insensitive parameters. This is in contrast to methods designed to rank parameters based on their sensitivities. There are many screening methods but the method of Morris is widely used and well suited for introducing the topic.

The goal of Morris screening is to gather information about how the direct sensitivities vary over the parameter space. This is a bridge between one-at-a-time sampling, which we have already criticized, and global methods where all parameters are varied. The idea is to repeatedly estimate $\frac{\partial QoI}{\partial p_i}$ in a systematic method and gather the statistics (in particular the mean and variance) of this over the parameter space. The method starts with a point in parameter space, $(\bar{p}_1, \bar{p}_2, ..., \bar{p}_n)$. The next step is to pick a random number between 1 and n – say i. This provides the next point in the Morris chain, $(\bar{p}_1, \bar{p}_2, ..., \bar{p}_i + \Delta p, ..., \bar{p}_n)$. The sensitivity in the ith direction at that point – referred to as the elementary effect, can be approximated by,

$$EE_i^1 \frac{QoI(\bar{p}_1, \bar{p}_2, ..., \bar{p}_n) - QoI(\bar{p}_1, \bar{p}_2, ..., \bar{p}_i + \Delta p, ..., \bar{p}_n)}{\Delta p}.$$

This is just the forward difference approximation of the derivative in the ith direction.

We repeat the procedure starting at the new point $(\bar{p}_1, \bar{p}_2, ..., \bar{p}_i + \Delta p, ..., \bar{p}_n)$. We collect all the elementary effects associated with the ith direction into an array,

$$\mathbf{EE}_k = (EE_k^1, EE_k^2, ..., EE_k^m),$$

where we have found m estimates of the elementary effect in the ith direction. We can then use the mean and standard deviation to measure the general landscape of the QoI. In this example, we generated a single chain. Part of the advantage of this method is the sampling structure is well-suited for parallelization by starting multiple chains.

The figures and calculations here were done using `SALib` which is a very flexible package. It does take some organization to structure the code – sample code is shown below.

```
from SALib.analyze import morris
from SALib.sample.morris import sample
from SALib.plotting.morris import
  ↪  horizontal_bar_plot, covariance_plot, \
    sample_histograms
problem = {
  'num_vars': 10,
  'names': ['X1', 'X2', 'X3', 'X4', 'X5',
    ↪  'X6', 'X7', 'X8', 'X9', 'X10'],
  'groups': None,
  'bounds': [[ 5.700e+00,  6.300e+00],
          [ 9.500e+01,  1.050e+02],
          [ 1.140e+02,  1.260e+02],
          [ 1.615e+02,  1.785e+02],
          [ 9.500e+02,  1.050e+03],
          [ 9.500e-02,  1.050e-01],
          [ 1.900e+00,  2.100e+00],
          [-1.050e+02,  9.500e+01],
          [ 3.800e+01,  4.200e+01],
          [ -6.300e+01,-5.700e+01]]
}
# Generate samples
param_values = sample(problem, N=1000,
  ↪  num_levels=4,  optimal_trajectories=None)
# Run the "model" -- this will happen offline
  ↪  for external models
tspan = np.linspace(0, 1, 100)
Y = np.empty([len(param_values)])
for i in np.arange(Num_samples):
    yp= solve_ivp(lambda t,Y:
      ↪  rhs(t,Y,param_values[i]),
      ↪  [tspan[0],tspan[-1]], [V0,m0,h0,n0],
      ↪  method='LSODA',t_eval=tspan)
    Y[i]=yp.y[0][-1] #value at the end of time
# Perform the sensitivity analysis using the
  ↪  model output
```

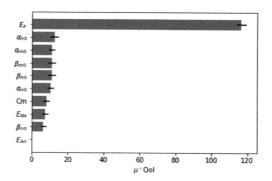

Figure 8.8: The Morris estimates for the sensitivities for the Noble model. The QoI is a measure of the frequency of oscillation.

```
Si = morris.analyze(problem, param_values, Y,
↪   conf_level=0.95,
                        print_to_console=True,
                        num_levels=4,
                        ↪  num_resamples=100)}
```

Note that Figure 8.8 shows that many have very low sensitivity with respect to our measure of the frequency. In particular, E_k plays a very outsize role while E_{An} does not alter our QoI at all. Once we have used a screening method to determine which parameters are insensitive, we can explore the behavior of the model as the parameters vary over the reduced parameter space. There is no theoretical cutoff for sensitivity but we will use a relative cutoff.

Figure 8.8 also indicates a difficulty with sensitivity. There are many parameters whose sensitivity index is approximately the same – but we have definitely learned that E_k matters a lot while E_{An} does not.

8.4   State of the Art and Caveats

The main distinction in this chapter is in the size of the parameter space. Screening methods are often used when the number of parameters is too large to use other methods. Similarly, they are useful if the individual realization of the models is computationally expensive. This happens when the models depend on time and space, or if there are large numbers of equations.

It is also important to note that Morris screening is both a global and local method. The goal is to estimate the local topography in parameter space (e.g. slopes); however, by varying all parameters as the steps are chosen, this information is collected in a global way. Screening methods are typically not used to asses uncertainty or robustness of prediction. Rather they are useful in identifying key parameters that should not be neglected or those that do not have a strong effect on the outcome. This identification can also provide insight into whether the parameters that may be estimated by comparing the model to data are "indentifiable". Parameters that are insensitive may be difficult to estimate since the values selected do not change the outcome, by definition.

8.5 Problems

Problems 8.1 *The idea of sampling in the unit cube requires taking the hypercube the parameters are chosen from, transforming each side into a unit interval, sampling on the unit cube and then returning it to the original space.*

(a) *Show that a parameter, p, whose range is $[p_1, p_2]$ can be transformed to the range $[0, 1]$ using a linear transformation. Specifically, consider the function $y(p) = a(p - p_1) + b(p - p_2)$ and determine a and b.*

(b) *Given a number in the unit interval, $\bar{y}$, determine the corresponding p value.*

Problems 8.2 *Specific physiological ranges of volumes are about 0.5 liters, with resistances on the order of 1–12 mm Hg/(liters/min) and compliance between 0 and 1. The number of heartbeats per minute at rest, F, ranges between 60 and 100.*

(a) *Show that the linear system defined in Equations 8.7, 8.8, and 8.9 has a unique solution for Q, P_a, and P_v.*

(b) *Use the fixed point code to find a solution to the nonlinear system in Equations 8.3, 8.4, and 8.5. Use the solution found in the previous part as the initial guess.*

(c) *Compare the histograms for $QoI_1 = P_v$ and $QoI = Q$ for different sampling methods. Does it appear that the sampling method alters the outcome dramatically? What about the time it takes to run the code? Make the number of samples larger? Does this change anything?*

Problems 8.3

(a) *Choose a different QoI and use the Morris method to sort the parameters into sensitive and insensitive groups.*

(b) *Compare the histogram of the full and reduced model.*

(c) *Explore methods for subtracting histograms in Python and/or MATLAB. Use this to quantify the difference between the full and reduced models.*

9

Neuroscience

Mathematical neuroscience is a well-developed field of mathematics. Mathematical theory has been used to develop drug targets and therapies. The Hodgkin-Huxley studies are fundamental to this theory. We also discuss the reduced Hodgkin-Huxley model referred to as the Fitzhugh-Nagumo model. One of the most used, variance-based sensitivity methods, Sobol' sensitivity, is discussed.

9.1   Historical Background

Along with ecology, modern physiology and specifically neuroscience was been essentially defined and structured around a fundamental theory that was developed to understand one striking set of observations. The history of neuroscience does not begin with Hodgkin and Huxley; however, their contribution cannot be overstated. By combining highly novel and refined experiments with elegant modeling allowed Hodgkin and Huxley to simultaneously propose, test, predict, validate, and refine hypotheses for the mechanism governing the electrical signaling in axons. The move from qualitative observations to quantitative measurements that informed theory that drove more insightful experiments bore full fruit in the collaboration between the two men.

Neurons consist of the cell body or soma, dendrites, and an axon. Broadly, the dendrites collect information, the soma process this and the axon send the processed data on to other neurons or cells. The information is encoded in electrical signals that are controlled by varying ion concentrations. As neurons are subjected to varying stimuli and specific conditions are met, the neuron can "fire" by exchanging large amounts of ions leading to a depolarization current that is transmitted

---

DOI: 10.1201/9781003316930-9

down the axon. The action potential is very similar to the waveform measured by an EKG that regulates the heartbeat see Figure 8.6.

Prior to their investigations, it was understood that axons carried some sort of information by an action potential that altered the permeability of the axonal membrane to ions. This action potential was assumed to be an all-or-nothing process that occurred simultaneously at all points along the axon and that the permeability of all ions were affected equally. Hodgkin had begun experiments that appeared to indicate that the action potential could be localized in specific regions of the membrane.

Hodgkin worked with K.S. Cole who, along with H.J. Curtis pioneered experimental electrophysiology. He recruited Huxley to join him at Cambridge in 1939. The team began experimenting on the squid giant axon [1]. These axons have a diameter of up to a millimeter so were large enough to manipulate. The basic experiment required sliding a wire down the length of the axon and recording the electrical activity.

Initial methodology was published in a short note in 1939 where one voltage trace was shown and the range of the maximum of the action potential was measured at 90 millivolts. The work was placed

---

Cole and Curtis

Kenneth Cole (1900–1984) and Howard Curtis (1906–1972) were the first to show how to change the conductivity in axons. Cole is often considered the father of biophysics. It was his voltage clamp method that allowed Hodgkin and Huxley to make controlled investigations of axon dynamics. The idea of the voltage clamp is to hold the membrane voltage at a fixed value by injecting a current calculated to cancel the residual voltage. This technique allows a patch of membrane to be held at a particular voltage so that the ionic flow can be measured. Cole learned this technique from his officemate George Marmont who never published and data using this method. Curtis collaborated with Cole on the original voltage clamp technique and moved on to radiation biology where he published more than 100 manuscripts.

---

[1]Not, as my advisor liked to joke, the giant squid axon

on hold during World War II and did not progress until 1945 where they published a more detailed manuscript providing a full explanation of their experiments. They also offered several potential mechanisms to describe the resting potential and the action potential measurements.

Two of these include electrical aspects of the membrane itself. First, rather than a barrier to the flow of ions, perhaps it has an additional capacitance that needs to be considered in series with the passive capacitance. The second possibility they considered was whether the membrane could act as an inductor and store and release voltage.

The other hypotheses discussed were whether the voltage could change the physical property of the membrane by reversing the orientation of the lipids that make up the membrane. Some estimates of the time scale provided support for this idea. Finally, they suggested that the permeability of the membrane could increase allowing ions in the exterior to enter the axon.

None of these were correct, although the first one provided some clue to the correct mechanism – permeability is affected but in a highly specific manner. Different molecules can be blocked or allowed to pass at different voltages. This line of thought lead directly to a theoretical model that was able to quantitatively capture the observations. More importantly, the analysis of the model provided specific predictions regarding the structure of the ion gates that allow ions to cross the membrane. This structure was confirmed in 2003 by Mackinnon who won the Nobel prize in chemistry for his work on visualizing the protein structure of the potassium channel.

The final piece of the story of the Hodgkin-Huxley model has to do with mathematical generalizations. We have stressed that quantitatively accurate models are needed to drive experiments by suggesting novel hypotheses, predict outcomes from interventions, and to refine existing theories. But another powerful use of mathematics is to strip away processes that dominate the smaller-scale attributes and focus on the main concept. For electrophysiological applications, the main objective is to understand the "excitability" of the process. Excitability has a technical description but we will use an intuitive description. Suppose that in the absence of any stimuli we consider a system at a basal state. Excitability means that a small stimulus in a limited region of time and/or space leads to a large deviation from the rest state and an eventual return to the rest state.

Action potentials are normal excitable behavior. A small stimuli – say the addition of some calcium ions – leads to a large change in the voltage before a return to the resting potential of a cell. The immune response to an acute challenge is also a normal, excitable reaction of the body to a foreign challenger. But what is the main mechanism that drives the deviation from rest? Is there a key feature that can be extracted that describes excitability?

In 1960 and 1962 Richard Fitzhugh was trying to understand how the structure of the action potential varied as different chemicals such as tetraethylammonium were added. He was exploring how the Hodgkin-Huxley model could be used to explain the action of the drugs on the gating variables and noted that by changing the rate at which the certain gating variables opened and closed altered the length and duration of the action potentials. This lead to a mathematical idea of time-scale analysis – essentially using quasi-steady-state analysis. In 1962, Fitzhugh derived what he referred to as the Bonhoeffer-Van der Pol equations which were a two-dimensional set of equations reduced from the four-dimensional Hodgkin-Huxley equations. The key behavior of excitability was recapitulated in these equations. The results are not quantitatively accurate to any particular experiment but were much easier to analyze and generalize the behavior.

Nagumo developed an analog circuit whose electrical properties could be modeled using the same equations that Fitzhugh proposed. This circuit was far easier to use to analyze the equations than the method Fitzhugh had been using – and was much more accurate. This allowed for a more refined exploration of the behavior. This, and subsequent developments, lead to attributing the equations to both Fitzhugh and Nagumo.

9.2 Action Potential

The equations described in Chapter 8 for the Noble model are essentially variations of the Hodgkin-Huxley equations. The main differences are how the gating variables are described. We will not attempt to give a comprehensive treatment of the gating variables but encourage the students to explore the primary manuscript from the Hodgkin-Huxley collaboration in 1952 [29, 30, 31]). It is an excellent example of blending theory and experiment.

The axon is treated as a capacitor which provides a relationship

---

Mechanical Calculators

It is fascinating to think about how the methods for determining numerical solutions to differential equations has evolved. Newton (1643–1727) and Euler (1707–1783) used tables to estimate the solutions to differential equations. One of the first methods for tabulating approximate solutions is "Forward Euler", which is easily implemented to solve differential equations of the form $\frac{dy}{dt} = f(y,t)$ approximating the derivative using the difference approximation, $\frac{dy}{dt} \approx \frac{y(t+\Delta t)-y(t)}{\Delta t}$. This provides an update method,

$$\frac{dy}{dt} = f(y),$$
$$\frac{y(t+\Delta t)-y(t)}{\Delta t} = f(y),$$
$$y(t+\Delta t) = \Delta t f(y) + y(t).$$

Therefore we can derive a sequence of estimates of the solution in time,

$$y(0) = y_0,$$
$$y(\Delta t) = y_1 = \Delta t f(y_0) + y_0,$$
$$y(2\Delta t) = y_2 = \Delta t f(y_1) + y_1,$$

$$\cdots$$

This is easily automated using a range of computer languages. But in the 1930's there were no computers. Hodgkin and Huxley used a Brunsviga mechanical calculator (see Figure 9.1).

Figure 9.1: A mechanical calculator of the sort used by Hodgkin and Huxley.

between the change in voltage and the current from Ohm's law,

$$C_m \frac{dV}{dt} = I.$$

A current is imposed from an external potential drop and is channeled into three paths representing the potassium, sodium, and leak currents. The circuit equivalent of the model is given in Figure 9.2.

9.2.1   Model: Hodgkin-Huxley

The current, driven by moving ions, can be broken down into different ions since the underlying hypotheses of Hodgkin and Huxley concerned the structure of the ion-specific membrane channels. A simple assumption is that the current is proportional to the difference in the current voltage and the voltage at rest. That is, with no stimulus, the membrane separates ionic separates ions from the intracellular space and maintains a specific gradient in each species. The proportionality is controlled by gating variables that depend on the voltage. The ions that are included are potassium and sodium and the others are included in a leak current that is analogous to the background current in the Noble model.

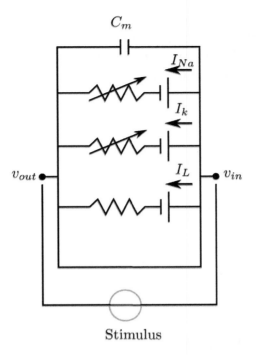

Figure 9.2: Circuit diagram of the membrane electrical activity based on the Hodgkin-Huxley equations.

Putting these together leads to the following system of equations,

$$C_m \frac{dV}{dt} = -\bar{g}_K n^4 (V - V_k) - \bar{g}_{Na} m^3 h (V - V_{Na}) \qquad (9.1)$$

$$-\bar{g}_l (V - V_l) + I_{app}, \qquad (9.2)$$

$$\frac{dm}{dt} = \alpha_m (1 - m) - \beta_m m, \qquad (9.3)$$

$$\frac{dn}{dt} = \alpha_n (1 - n) - \beta_n n, \qquad (9.4)$$

$$\frac{dh}{dt} = \alpha_h (1 - h) - \beta_h h, \qquad (9.5)$$

The term I_{app} refers to a current that can be applied through a stimulus – for example via electrodes in voltage-clamp experiments.

The coefficients of the gating variables can be thought of as the frequency of finding closed gates $(1 - m)$ and open gates (m). The rate that the gates transition between open and closed are encoded in the α and β coefficients. These were found empirically by fitting specific experiments so have a very nonintuitive feel,

$$\alpha_m = 0.1\frac{25 - v}{e^{\frac{25-v}{10}} - 1}, \tag{9.6}$$

$$\beta_m = 4e^{-\frac{v}{18}}, \tag{9.7}$$

$$\alpha_h = 0.07e^{-\frac{v}{20}}, \tag{9.8}$$

$$\beta_h = \frac{1}{e^{\frac{30-v}{10}} + 1}, \tag{9.9}$$

$$\alpha_n = 0.01\frac{10 - v}{e^{\frac{10-v}{10}} - 1}, \tag{9.10}$$

$$\beta_n = 0.125e^{-\frac{v}{80}}. \tag{9.11}$$

9.2.2 Analysis

We can now explore the excitability by adding a small deviation from steady-state voltage (this mimics adding a pulse of current). Notice that if the applied current is small there is a small excursion that does not have the hallmarks of an action potential. In particular there is no plateau or rapid transition to the hyperpolarized state. As the applied current is increased, we see the evolution to an action potential.

By maintaining the applied current at a constant level can also lead to a different physiological mechanism. For example, Hodgkin and Huxley demonstrated that there is a refractory period after an action potential where additional stimulation does not lead to an additional action potential. This can be observed using numerical solutions as in Figure 9.3. The model can also show tonic firing.

9.2.3 Sensitivity Analysis: ANOVA – Sobol'

In this section, we introduce one of the most common methods to assess sensitivity. The method is based on the ANOVA (Analysis of Variance) and partitions the variance between the input parameters. This method was introduced by I.M. Sobol' in 1993 [32]. We will describe

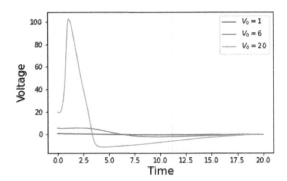

Figure 9.3: Comparison of Hodgkin-Huxley trajectories for different initial voltage perturbations. Note that there is a cut-off, below which no action potential propagates.

the derivation below; however it is interesting to discuss what makes this such a popular method for assessing sensitivity. There are at least two distinguishing features of Sobol'. First, this method applies to non-linear relationships. Recall that most methods including linear regression, PRCC and scatterplots require some sort of linearity to provide a single value to a parameters sensitivity. Sobol' does not require this. Second, Sobol' provides information about the interactions between parameters. PRCC, which is arguably the other most used sensitivity method, *discounts* these interactions and only considers the primary relationship between the input parameter and the QoI [57].

It is important to understand that the underlying framework for Sobol' estimates is quite different than what we have explored until now. Most of the previous methods can be understood graphically or geometrically. One can think of the input/output relationship in a spatial sense and get some understanding of what sensitivity means. Sobol' sensitivity starts with understanding of the variance of an input/output relationship. Relating the sensitivity to the variance is a different outlook than correlation. A QoI is correlated with a parameter if changes in the parameter cause changes in the QoI. A QoI has high variance due to a parameter if changes in the parameter induce more change relative to the mean – that is the QoI has more spread about the mean [57].

Analysis of variation (ANOVA) is a widely used method in statistics quantify variations in the means of different groups. It was first introduced by Fisher to quantify the relationship between twelve different potato varieties and three fertilizers on the crop yield [18]. Because there are two parameters (type of potato and fertilizer) and one QoI (crop yield), we have to do something that considers the interactions since it may be that the fertilizer/potato combination matters. Fisher did this by showing that the total variation can be decomposed into the sum of the variance in means within each group (species of potato) and the variance in means between each group (variance in means between potato species and fertilizer type). The underlying insight from Fisher was that this comparison can be used to determine whether specific variations have a measurable impact on specific measurements. From a statistical standpoint, Fisher developed a method to test hypotheses (like whether the null hypothesis that means of the potato group and the fertilizer groups are equal is true).

Anova and Potatoes[2] Consider the yield of four potato varieties with three different fertilizer levels, $Y(V,F)$. After one season the yields are,

$$
\begin{array}{ccccc}
\text{Yield} & V_1 & V_2 & V_3 & V_4 \\
F_1 & \lceil 109.0 & 110.9 & 94.2 & 125.9 \rceil \\
F_2 & \mid 104.9 & 113.4 & 110.1 & 138.0 \mid \\
F_3 & \lfloor 151.8 & 160.9 & 111.9 & 145.0 \rfloor
\end{array}
$$

The average yield, $\bar{Y}$, is 123.0 across all harvests. We can decompose the yield matrix into the sum of the total mean (123.0), and the main effects which are the variances in response due to fertilizer and variance due to variety. The main effect helps answer whether one particular fertilizer or variety raised or lowered the yield.

The variance in yield due to fertilizer is i, $\frac{1}{3}\sum_{j=1}^{3}(Y_{ij}-\bar{Y})$,

$$
\begin{array}{ccc}
F_1 & F_2 & F_3 \\
-13.0 & -6.4 & 19.4
\end{array}
$$

---

[2]based on a talk by Art Owen and the paper by Fisher (1923)

For the variety we find,

$$
\begin{array}{cccc}
V_1 & V_2 & V_3 & V_4 \\
-1.1 & 5.4 & -17.6 & 13.3
\end{array}
$$

We can use these to decompose each of the yields into variances due to fertilizer type, potato type, and the rest (interaction terms):

$$
\begin{array}{ccccccccc}
\text{Yield} & = & \text{Mean} & + & \text{Fertilizer} & + & \text{Type} & + & \text{Interactions} \\
109.0 & = & 123 & - & 13.0 & - & 1.1 & + & 0.1
\end{array}
$$

Thus, this fertilizer and type choice tends to lower the yield with the fertilizer choice having more effect. Also, the combination of type and fertilizer did not have a strong combined effect.

Sobol generalized this and showed that any function can be decomposed into the sum of specific, orthogonal functions that make it very easy to derive a different decomposition of the total variance. This is similar in spirit to ANOVA where the 'function' is a discrete grid of QoI in terms of parameters. Sobol' showed that for a very general functional relationship between Y and parameters X_i, $Y = f(X_1, X_2, ...X_p)$ there is a unique sequence of functions that can be used to construct Y that have two properties. First, the functions have specific interdependencies between the parameters and second that they are orthogonal. The first means that we can write,

$$
\begin{aligned}
Y = f(X_1, X_2, ...X_p) & = f_0 + \sum_{1}^{p} f_i(X_i) + \sum_{1 \leq i < j \leq p} f_{ij}(X_i, X_j) + ... \\
& = + f_{1,...,p}(X_1, ..., X_p).
\end{aligned} \tag{9.12}
$$

The second property of orthogonality allows us to use Equation 9.12 to relate the total variance to the variance restricted to parameter combinations. The orthogonality can be written,

$$
\int_{[0,1]^p} f_{i_1,...,i_p}(X_{i_1}, ..., X_{i_q}) f_{j_1,...,j_p}(X_{j_1}, ..., X_{j_q}) d\mathbf{x} = \mathbf{0}. \tag{9.13}
$$

This is a notationally dense way to say that the integral of any of the functions that are used to construct Y against any other is zero.

By squaring both sides of Equation 9.12 and using the orthogonality, we find a relationship between the total variance and variance due to the parameters,

$$\int_{[0,1]^p} f^2 d\mathbf{X} = \mathbf{V} = \sum_{i=1}^{p} V_i + \sum_{i<j} V_{ij} + \sum_{i<j<k} V_{i,j,k}$$
$$+ \dots + V_{1,2,3,\dots p}, \tag{9.14}$$

where

$$V_{i_1,i_2,\dots,i_s} = \int_{[0,1]^p} f^2_{i_1,i_2,\dots,i_s} dX_{i_1}, \dots, dX_{i_s}, \tag{9.15}$$

is the variance due to the specific parameter combinations $(X_{i_1}, \dots, X_{i_s})$.

It is more useful to consider the comparisons of the partial variances to the total variance, so that Equation 9.14 can be written as,

$$1 = \frac{\sum_{i=1}^{p} V_i}{V} + \frac{\sum_{i<j} V_{ij}}{V} \dots$$
$$+ \frac{\sum_{i<j<k} V_{i,j,k} + \dots + V_{1,2,3,\dots p}}{V}, \tag{9.16}$$

by dividing by V.

The *main or first order effect* of parameter X_i is just $\frac{V_i}{V}$. This provides a measure of how much variation in a single parameter varies the QoI compares to that of varying all parameters, neglecting any interactions.

A related measure is the *total effect* of parameter X_i. Here, interactions are also included. Each of the terms that includes the parameters X_i,

$$S_{T_i} = \frac{V_i}{V} + \frac{\sum_{i \neq j} V_{ij}}{V} \dots$$
$$+ \frac{\sum i \neq j, k V_{i,j,k} + \dots + V_{1,2,3,\dots p}}{V}. \tag{9.17}$$

There are many ways to interpret the main and total effects. One can think of these as how much knowledge of the exact value of a single parameter, X_i, reduces the variance in the QoI, where the main and total provide under-and overestimates of the effects. For us, we will not go too far in this direction but note that these provide rankings of the parameter sensitivities and are applicable even if the QoI depends nonlinearly on the parameters.

One other note that is important to keep track of. Under the hood of the different implementations of Sobol' measures are integral estimates. It is difficult to see here, but the process of moving from the Sobol' decomposition of the QoI to the variance requires multidimensional integration – where the dimensionality is given by the number of parameters. We are right back to the curse of high dimension. Many implementations of Sobol' use Monte Carlo integration and others use use different sampling algorithms. One of the most used is FAST [46] or eFAST [56] which use a spectral method similar in spirit to Fourier decompositions. Again, we will not go into the details but from a practical standpoint, sampling must be done with care. It is extremely important to ensure that enough samples that the integral estimates have converged. There is no fixed number – it depends on the parameters, dimensionality, model and other properties. Many packages have converge criteria but we should be mindful that Sobol' requires orders of magnitude more samples that other sensitivity methods such as PRCC to get decent estimates. There are trade-offs between flexibility and efficiency.

There are many, many ways to calculate the Sobol' indices. We will use a relatively robust method based on Sobol's original algorithm. This is implemented in `SAlib` in Python. MATLAB has multiple versions that can be implemented and we provide examples along with our codes.

We can then think of the QoI for the Hodgkin-Huxley model. For us, a simple measure might be the maximum of the action potential. This can provide some insight into parameters that can make the system more excitable – that is which parameters can change whether applying a current leads to a small excursion from rest or a large excursion from rest. See Figure 9.4 for the main effects.

9.3 Fitzhugh-Nagumo

As we have discussed, the Hodgkin-Huxley model does a great job in providing a framework to develop highly detailed physiological

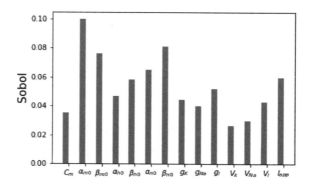

Figure 9.4: Main effects for the sensitivity of the Hodgkin-Huxley equations the QoI equal to the maximum voltage.

models. But it does make it more difficult to see the governing process for the excitability. The Fitzhugh-Nagumo provides an elegant connection between the physiologically based models and excitability.

9.3.1   Model

The reduction of Hodgkin-Huxley to a two variable model uses a quasi-steady-state assumption. Recall that this is an approximation that treats one equation as a "fast" variable compared to the others. This implies that we can neglect the dynamic change and assume this happens instantly. For Hodgkin-Huxley, the m-gate equilibrates quickly so we assume that $\frac{dm}{dt} \approx 0$. Thus $m \rightarrow$ a function only of v, $m_\infty(V)$. Similarly, h can be approximated by a linear shift of n. After quite a bit of simplification we can reduce the model to,

$$C_m \frac{dV}{dt} = -\bar{g}_K n^4 (V - V_k) - \bar{g}_{Na} m_\infty^3 (0.8 - n)(V - V_{Na}) \quad (9.18)$$

$$-\bar{g}_l(V - V_l) + I_{app}, \quad (9.19)$$

$$\frac{dn}{dt} = \alpha_n(1 - n) - \beta_n n. \quad (9.20)$$

This is still a bit unwieldy, but looking at the right-hand-side of Equation 9.18, we see that the rate of change of the voltage is related to the voltage which is embedded in a cubic function. The right-hand-side of the gating variable is linearly related to the voltage. This can be reduced to a simpler looking expression that is related to the van der Pol oscillator,

$$\varepsilon \frac{dv}{dt} = f(v) - w - w_0, \tag{9.21}$$

$$\frac{dw}{dt} = v - \gamma w. \tag{9.22}$$

The function $f(v) = Av(v - \alpha)(1 - v)$ relates the current to the voltage and w plays the role of the gating variable.

9.3.2 Analysis

The Fitzhugh-Nagumo equations are two-dimensional making geometric analysis relatively straight-forward. The main consideration is the null-clines steady-states and dependence on the imposed current. Looking at Figure 9.5, we see that the linear gating variable null-cline can intersect the cubic, voltage null-cline in several ways – altering the dynamics.

There can be from one to three intersections (e.g. fixed points). The most interesting ones, from the standpoint of relating this to Hodgkin-Huxley are the single steady-state with low voltage and the single steady-state with an intermediate value. The former is primed for excitability – there is a low resting state. With an impulse of added current, there is a large excursion (e.g. an action potential) (see Figure 9.6).

The case where there is a single, unstable equilibria corresponds to the case of a constant application of a current. When Hodgkin and Huxley performed this experiment, it led to tonic firing. It can be shown that there is a limit-cycle (e.g. a periodic solution that attracts the trajectories) in this case (see Figure 9.7).

9.3.3 Sensitivity: Moment Independent

Sobol' sensitivity focuses on the variance of the output distribution of the QoI and apportioning this among the input parameters. This can be described in terms of central moments of the distribution since the variance is the second moment. Recall here that the term "moment" refers

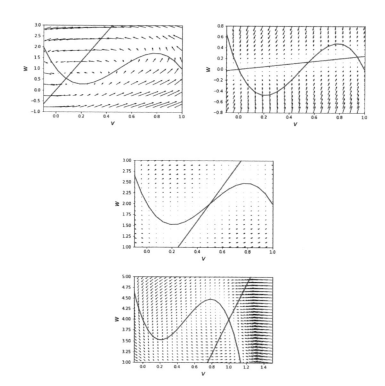

Figure 9.5: Nullclines for different parameter regimes demonstrating examples of one steady-state (both stable and unstable) and three steady-states (bistable case).

to specific integrals of the function that describe its graph. Moments can be related to forces in statics (which is where most calculus students encounter these), descriptive statistics (heavy-tailed, symmetric, skew, etc.), and for other building blocks describing functions.

It is far too much to go into detail here, but one observation that does matter is that correlations between parameters can alter the moment calculations. This means that Sobol' measures may be incorrect for correlated parameters. In practice this is typically not an issue since most true parameters in models should be thought of as independent. Dependencies might arise, for example viscosity depends on the temperature. However, most models either formally neglect this (assuming the temperature does not vary very much) or provide a closure model

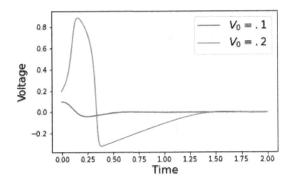

Figure 9.6: The voltage trace for the excitable steady-state. The parameters are: $\varepsilon = 0.01$, $A = 1$, $\alpha = .1$, $w_0 = 0$, and $\gamma = .5$. Here there is one stable steady-state $(V, w) = (0, 0)$. With a small perturbation in voltage, there is a small deviation. However, with a bit more voltage input, a large excursion occurs reminiscent of an action potential. This is referred to as excitability.

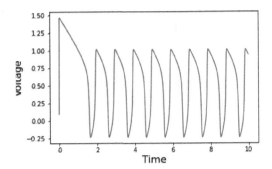

Figure 9.7: Using the same parameters as in the case of an excitable steady-state, with the exception of w_0, which is -1 in this case. We see tonic firing, with self-sustained oscillations.

that connects the two. These models are often referred to as empirical models and depend on other parameters.

Regardless of this, other sensitivity measures have been developed to address different geometric descriptions of the distributions. One of the most popular is referred to as the *moment independent measure*

$= \delta_{X_i}$. This has the property that if the output distribution is independent of a parameter $\delta(X_i) = 0$. If all parameters are considered $\delta = 1$. Detailed understanding is beyond this textbook but with some elementary statistics, δ-measures can be related to expectation of different conditional probabilities.

We will again, show the implementation in `Salib`, so the syntax is very similar to the previous examples in this section. We introduce a new QoI, however. One question that we have not addressed is how sensitivity can help with fitting models to data. There are many ways to *parameterize* models – that is there are many ways to determine parameters that reduce the error/discrepancy between the model and observations. There are methods that do this by exploring parameter space in some guided method. In Matalb `fminsearch` is often used while Python users may use the packages from `SciPy`, in particular the genetic evolution method is very robust. If your parameter space is large it may be quite difficult to determine your parameters. Even worse, there are some parameters that may not be identifiable from the data – that is, the data may not provide enough information for the methods to determine the parameters uniquely. The latter is a ubiquitous issue in the sciences because all of our predictions rest on parameters and if we are not certain about the values, we may make large errors in predictions.

Sensitivity can help distinguish which parameters can be well-estimated using specific data sets. Given data at discrete time-points, t_i, we define,

$$\begin{aligned} QoI &= \|y_{data} - y_{model}\|_2 \\ &= \sqrt{\sum_i \left(y_{data}(t_i) - y_{model}(t_i)\right)^2}. \end{aligned}$$

We then determine which parameters are most significant for this QoI. We have higher confidence in our estimates of these parameters, since errors in these estimates would change the QoI much more.

We use Moment Independent measures and given data for the Fitzhugh-Nagumo model. The data are not from specific observations in the lab. Instead, we take a known set of parameters, numerically simulate the differential equations. We then add some noise to these and

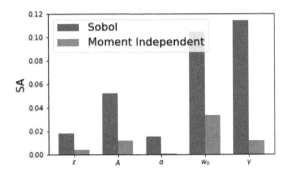

Figure 9.8: Comparison between Sobol ranking and moment independent measures. We see that ε and α play a less significant role in the discrepancy between the model and the data.

refer to this as "simulated" data. We can then calculate the sensitivity and show what happens if we change the insensitive parameters.

In Figure 9.8, we show two sensitivity measures – first Sobol and second the moment independent measure. We see that these agree, qualitatively. Because of the definition of the measures there is no reason to expect quantitative agreement. In particular, all variance-based measures are relative to the variance, there can be wide differences in the magnitudes between measures. However, we see from the figure that both methods agree that there are two parameters that are far less significant than the other three parameters.

In Figure 9.9, we compare the voltage traces for different values of the insignificant parameters. We also show voltage traces for varying the significant parameters. This shows how this sensitivity ranking helps indicate which parameters we got right for the fitting. We note that comparing the rankings and the qualitative differences, variation in α from 0.1 to 1 leads to the smallest difference, followed by ε as it is varied from .01 to .1 and varying w_0 by a much smaller amount, from -1 to -1.5, drastically changes the solution.

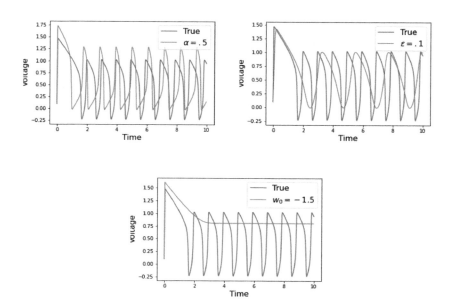

Figure 9.9: Comparison between the solutions using the true parameters and varied parameters. Notice the difference between varying insensitive parameters by one order of magnitude versus varying a significant parameter by one half an order of magnitude.

9.4   State of the Art and Caveats

Sobol' indices are one of the most often used sensitivity methods. Along with PRCC, Sobol' is widely accepted. There are numerous implementations in a variety of platforms. Sobol' has the benefit of not requiring monotonicity in the parameter/QoI relationships. This makes it very attractive since we do not need to look at scatterplots. However, in exchange for this flexibility, Sobol' typically requires many more samples to converge. In essence, Sobol' requires the estimation of multivariable integrals – often using Monte Carlo methods. To accurately estimate these integrals may require orders of magnitude more samples than PRCC.

One fo the difficulties in most implementations of Sobol' is determining if the estimations have converged. In most codes, we have to set the number of samples at the beginning. Because of the underlying structure of the Sobol' decomposition, we can-not add more samples if we find the estimates have not converged. Doing this typically impacts the estimates making them highly suspect. So, if the estimates are changing, it is typically required that the code is re-run for more samples. There are efforts to do this within the codes and likely this will be become less of an issue in the future. For now, it is likely more useful to take more samples rather than less and be patient while waiting for the results.

9.5 Problems

Problems 9.1 *(a) For the example about the potato experiments, which fertilizer has the strongest effect on the yield? Discuss the difference between fields $(3,1)$ and $(3,4)$.*

(b) Write the entire yield matrix as the sum of a matrix with rows equal to the main effects due to fertilizer, a matrix with columns equal to the main effects due to potato type and a matrix with the differences.

(c) Show that the rows of the two main effect matrices are orthogonal (e.g. the dot product is zero) and that the columns are orthogonal.

Problems 9.2 *(a) Compare the qualitative behavior for the Noble model and the Hodgkin-Huxley Model.*

(b) Use the $QoI = \int_t V_{Noble} - V_{Hodgkin-Huxley}$ and determine which parameters are most responsible for differences by using the Sobol' indices for the combined model. The workflow will require sampling over parameters for both the Noble and Hodgkin-Huxley model, determining the differences and using Sobol' decomposition to estimate the indices.

Problems 9.3 *Consider the forced Van Der Pol Oscillator,*

$$\frac{dx}{dt} = \varepsilon \left(x - \frac{1}{3}x^3 - y \right) \tag{9.23}$$

$$\frac{dy}{dt} = \frac{x}{\varepsilon}. \tag{9.24}$$

(a) Sketch the null-clines and the behavior in the phase-plane.

(b) Use numerical methods to explore what happens for large ε? For small ε?

(c) Consider the forced equations,

$$\frac{dx}{dt} = \varepsilon\left(x - \frac{1}{3}x^3 - y\right) + F\cos\left(\frac{2\pi t}{T_{in}}\right) \quad (9.25)$$

$$\frac{dy}{dt} = \frac{x}{\varepsilon}. \quad (9.26)$$

Explore the sensitivities for small and large ε and small and large T_{in}.

Problems 9.4 *Consider the Hodgkin-Huxley model where all but three parameters are fixed at their nominal values. Find a set of parameters where you can demonstrate that the total effects may sum to more than one. Looking at the Sobol' expansion for exactly three parameters, explain how the main effects can sum to more than one. Is it possible to determine which parameter controls this?*

Problems 9.5 *One difficulty with Sobol is how many samples are needed. Explore the behavior of the indices for the Hodgkin-Huxley equations while varying the number of samples. What happens if the number of samples is too small? Examine the difference between total indices and the main effects – what does this tell you?*

10

Genetics

Genetics has undergone tremendous expansion in the past decades. The amount of knowledge unlocked by the human genome project and techniques has rapidly changed the impact of theory on predictions leading to personalized medicine and drug development. We use the Factorial Design sensitivity which is an example of a brute force method that is related to experimental design.

10.1   Historical Background

How are specific traits passed between generations? What is the link between the microscopic controls built into our genetic structure and variations within species over time, distance, or environment? What is the role in mathematics in describing species variations and establishing predictive and descriptive hypotheses?

Genetics and heredity has gone from a subject based on rule-of-thumb, practical observations leading to selective breeding, to a highly specialized field requiring massive amounts of computer time to process the data available and capable of identifying potential diseases in children who are not even conceived.

The hinge experiments were performed by a patient, thorough monk examining pea plants over the course of 8 years. Prior to Gregor Mendel, little had changed since Aristotilian philosophers speculated that all traits were passed down the generations by sharing the building blocks for each part of an organism. The kidney passes very small "kidney" blocks while the heart passes small "heart" blocks. Each can be mixed between father and mother to build the offsprings kidney and heart. This helped explain how selective breeding can change specific characteristics by providing more similar blocks to build with. It also

DOI: 10.1201/9781003316930-10

explained why some children look more like their mother or father. On the surface it is not incompatible with observations made through the 19th century.

As Mendel used controlled observations of pea plants, he was able to demonstrate that offspring are not a blend of the progenitors. When he controlled the pollination and crossed two plants with distinct traits, the offspring showed one trait and not a blend of the traits. For example, crossing a plant that produced only yellow peas with a plant that bred true to produce only green peas did not lead to a plant that produced a yellow/green mixture. This implies that what is passed down is far more discrete than pangenesis theory assumed.

Even more importantly, by systematic studies, Mendel was able to quantify the rate of propagation and make quantifiable predictions. Developing a theoretical framework, Mendel completely altered the understanding of genetic inheritance. He introduced the concept of dominant and recessive traits and was able to verify his theoretical framework through highly refined experiments.

More and more attention was paid to the consequences of Mendel's theory. This required more sophisticated mathematics – especially when the link between statistics and genetics was clarified in the late 1800's. This required more sophisticated mathematics and lead to various predictions, discrepancies and confusion. Because of the personalities of the particular scientists, there is actually quite a lot of biographical and autobiographical discussion concerning one of the core quantifiable results regarding genetics. These equations, referred to as the Hardy-Weinberg equations that are used to provide an estimate for the distribution of genes within a population at steady-state.

A leading question was posed to Hardy, by Punnett [55] in 1908. Punnett (of the Punnett square) was a well known supporter of Mendel's theory of heredity and was already well known at the time. He was asked why there are still blue-eyed people even though brown is the dominant trait. As Punnett said: "Knowing that Hardy had not the slightest interest in genetics I put my problem to him as a mathematical one. He replied that it was quite simple and soon handed to me the now well-known formula $pr = q^2$" [55].

Hardy was a highly respected mathematician and devoted to pure mathematics. To say that he was not motivated by genetics is an understatement. Hardy famously said [25], "I have never done any-thing *useful*. No discovery of mine has made, or is likely to make, directly or indirectly, for good or ill, the least difference to the amenity of the world. I have helped to train other mathematicians, but mathe-maticians of the same kind as myself, and their work has been, so far at any rate as I have helped them to it, as useless as my own. Judged by all practical standards, the value of my mathematical life is nil; and outside mathematics it is trivial anyhow. I have just one chance of es-caping a verdict of complete triviality, that I may be judged to have created something worth creating. And that I have created something is undeniable: the question is about its value".

This is often considered a mathematicians credo and it is interest-ing that the Hardy-Weinberg equation provided an elegant answer (and more importantly a framework for extending the answer) to a long-standing open question in genetics. This, in turn, led directly to a mas-sive expansion in the understanding of heredity – leading to changes in crop and animal breeding science, medical advances that push well into the modern era of treatments for HIV, COVID, and other diseases.

The next major milestone was about 10 years later when Fisher connected genetics with evolutionary biology. His fascination with Darwin's and Wallace's theory of evolution and natural selection. Fisher had major contributions to multiple scientific areas including statistics – where he introduced the concept of variance – genetics, and measures that were later adapted for sensitivity quantification. Fisher is one of the towering figures of science. Unfortunately, his scientific endeavors were overshadowed by his stance on racial traits, eugenics, and heredity. He wrote formal support of the Nazis after World War II. It is difficult to find someone who played a pivotal role in establishing a field of science that benefitted society at the same time as supporting one of the great criminal tragedies. There was also rapid progression in understanding how genetic information can be encoded. These in-clude Rosalind Franklin's x-ray diffraction that lead directly to Watson and Crick's insight into the structure of DNA. From there, Sanger and Gilbert were able to perform the first DNA sequencing in the 1970's leading to the Noble price in 1980 and to incredible changes in how diseases are diagnosed, treated, and specific drugs developed.

An indirect effect of the experimental transformation is the massive amount of data that must be processed. The human genome project generated an unprecedented amount of data with current estimates of

---

Fisher 1890–1962

Fisher is an example of a very difficult scientist to focus on. He clearly made tremendous contributions to many fields – he is credited with much of the framework of modern statistics based on work he did at Rothamsted Experimental Station from 1919–1933. ANOVA is one of the most widely used methods in statistics and systematically studied population genetics developing the methodology that is still used.

At the same time, he was a difficult person to interact with. He held a long-running and bitter feud with Karl Pearson who was the Galton Professor of Eugetics at University College. The feud apparently started when Pearson criticized one of Fishers submitted papers. Fisher was furious and when Pearson continued to attack Fishers scientific contributions, the feud became very public and Fisher continued to attack Pearson even after Pearson's death.

He was also well-known to support racial supremacy arguments and eugenics. He wrote supportive testimony for the Nazi Otmar Freiherr von Verschuer who was at Auschwitz arguing that his scientific contributions overshadowed his support of Hitler. A similar argument put Fisher at odds with the United Nations Education, Scientific and Cultural Organization (UNESCO) when the organization attempted to grapple with the definition of *race* arguing that there was strong genetic predisposition for "intellectual and emotional development".

By scientific merits, Fisher is a major figure. The debate about how to approach his contributions is ongoing but not simple – it is not possible to just remove him and his science. It is also not reasonable to ignore the context of his scientific endeavors. It is part of our responsibility to try and separate the science and the portions that are (T)rue from the parts that are not quantifiable.

2–40 exobytes (an exobyte is close to 1 million times more data than a typical home computer) per year. Dealing with data requires specific tools and methods and it is interesting to note that sensitivity plays a significant role in the way forward. It is impossible to blindly process all of this data without knowing which parts of the data provide information for specific questions – that is sensitivity with respect to a QoI.

10.2 Heredity

Hardy-Weinberg equations are used to calculate the equilibrium proportions of genetic expression based on several simplifications. The simplifications include no mutations, no natural selection, and random mating. Under these simplified assumptions, it is relatively simple to determine the genetic distributions and allele frequencies. The calculations were suggested by Punnett and, the simplest versions, look very similar to Punnett square calculations. Consider a single allele with a dominant and recessive trait – A and a. Hardy-Weinberg equations consider the frequency of each trait in the population, p and q and the predicted frequency in the next generation. The next generation frequency can be found by calculating p^2 (the frequency of homozygote dominate), q^2 (homozygote recessive), and $2pq$ (frequency of heterozygote dominant). since these are frequencies $p + q = 1$ and $p^2 + 2pq + q^2 = 1$ and p and q are determined. It is simple to see that after at most one generation, the frequencies are unchanged.

But what if genetic drift, mutations, and natural selection are included? This becomes much more challenging. One needs some sort of model for the dynamics that insert, remove or alter the alleles. One of the most useful steps in this direction was given by Fisher in 1930 [17] when he stated that "The rate of increase in fitness of any organism at any time is equal to its genetic variance in fitness at that time". This is often referred to as the fundamental theorem of natural selection. This relates the state of an organism to the previous state. However, it is quite an imprecise statement and has led to many different mathematical interpretations. In the next section, we connect Fishers model with an equation referred to as a replicator equation which has direct connections to game theory. Interestingly, game theory was developed by John Nash to study specific economic principles – indicating how portable mathematics can be. From a monk studying pea plants, to one of the most celebrated pure mathematicians in history, to a celebrated economist connected by common mathematical structures.

10.2.1 Mathematics

One can interpret Fishers fundamental theorem of natural selection as a statement that relates the rate of change of a population (and the genetic expression) and the fitness of the population that also relates to the genetic structure [3]. We could write this as a differential equation for the ith genotype in a population – say P_i,

$$\frac{dP_i}{dt} = f_i(P_1, P_2, ..., P_n)P_i. \tag{10.1}$$

Where f_i is the fitness landscape – essentially a measure of the reproductive rate of P_i. The mean fitness is,

$$\bar{f} = \frac{\sum_{j=1}^{n} f_j(P_1, P_2, ..., P_n)P_j}{\sum_{j=1}^{N} P_j}, \tag{10.2}$$

$$= \sum_{j=1}^{n} f_j(P_1, P_2, ..., P_n)p_j, \tag{10.3}$$

where p_i is the fraction of the population that is in the ith genotypic state.

the variance of the fitness is,

$$\text{Var}(f) = \sum_i p_i(f_i - \bar{f})^2.$$

With a little work we can rewrite Equation 10.1 in terms of mean fitness and the population fractions. We begin by looking at the derivative of the replicator fraction,

$$\frac{dp_i}{dt} = \frac{d}{dt}\frac{P_i}{\sum P_i},$$

$$= \frac{\frac{dP_i}{dt}}{\sum P_j} - \frac{P_i\left(\sum_j \frac{dP_j}{dt}\right)}{\left(\sum_j P_j\right)^2}.$$

This leads directly to,

$$\frac{dp_i}{dt} = (f_i - \bar{f})p_i. \tag{10.4}$$

This equation is referred to as the replicator equation. Notice that p_i increases if the local fitness landscape is greater than the average fitness. For a specific genotype to propagate it just needs to be more fit than the average population.

As an example, we use a model to describe the variation over time of the pattern of spots on a species of lizards [60]. Males with orange throats are very aggressive and defend large territories. Males with dark blue throats are less aggressive and defend smaller territories. Here, there are males with either orange, yellow, or blue throats – each color has a specific method of defend territory. There are "sneakers" who do not defend any territory, aggressors who defend large territories and lizards that are in the middle and defend smaller territories and are less aggressive.

10.2.2 Analysis

The differential equations for rock-paper-scissors can be written as,

$$\frac{dp_r}{dt} = p_r(1 - p_r - \alpha_r p_p - \beta_r p_s), \tag{10.5}$$

$$\frac{dp_p}{dt} = p_p(1 - p_p - \alpha_2 p_s - \beta_2 p_r), \tag{10.6}$$

$$\frac{dp_s}{dt} = p_s(1 - p_s - \alpha_3 p_r - \beta_3 p_p), \tag{10.7}$$

As written this is referred to as the Leonard-May equation. There are major differences between the symmetric case – when the $\alpha_1 = \alpha_2 = \alpha_3$ and $\beta_1 = \beta_2 = \beta_3$– and the asymmetric case. For the symmetric case, it is relatively straightforward to verify that there are four equilibria: $(1,0,0)$, $(0,1,0)$, and $(0,0,1)$ and an interior steady-state: $\frac{1}{1+\beta+\alpha}(1,1,1)$. Under suitable conditions on α and β, the interior is a saddle point. Numerical behavior is very interesting with increasingly long times where one species dominates with rapid transitions to another dominant species (see Figure 10.1).

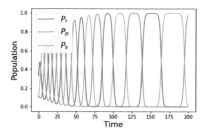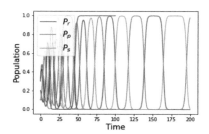

Figure 10.1: Comparison between the solution of the rock-paper-scissors model with different, symmetric parameters. For the first $\alpha_1 = \alpha_2 = \alpha_3 = .2$ and $\beta_1 = \beta_2 = \beta_3 = 2$. The second is $\alpha_1 = \alpha_2 = \alpha_3 = .05$ and $\beta_1 = \beta_2 = \beta_3 = 3$.

10.2.3 Sensitivity: Factorial Design

Another method for sensitivity quantification is based on the *design of experiments* framework [13]. What is an efficient method for designing experiments to test variations in different controllable parameters? Rather than randomly changing parameters and combinations of parameters, it is better to have a structured method to vary inputs, measure outputs, and quantify the response.

In this chapter, we will focus on the simplest case referred to as the 2-factor, full factorial design. Here we vary input parameters between a high and low value. We take all possible combinations of parameters and values and average the difference of results for high and low values. This can be arranged as in Table 10.1 where $-$ indicates a low value and $+$ indicates a high value. The averages are referred to as the main effects and can be directly related to linear regression. This means that nonlinear effects and interactions are neglected. There are many extensions to this – in fact SALib uses a fractional factorial method [28]. We will show a simpler method to estimate the main effects.

One can also add some stochasticity by selecting a nominal parameter from a distribution and re-running the entire procedure n-times. This leads to n main effects which can be used to provide descriptive statistics (mean and standard deviations or confidence intervals).

Table 10.1: Factorial design table. The symbols indicate using the lower or higher value of the parameter. Each run is defined by a parameter set given by each row of the table.

Run	Parameter 1	Parameter 2	Parameter 3	QoI
1	-	-	-	QoI 1
2	+	-	-	QoI 2
3	-	+	-	QoI 3
4	+	+	-	QoI 4
5	-	-	+	QoI 5
6	+	-	+	QoI 6
7	-	+	+	QoI 7
8	+	+	+	QoI 8

Note that the sign of the main effects have the same interpretation as in PRCC – a negative (positive) sign indicates that increasing the parameter leads to a decrease (increase) in QoI.

To demonstrate this using the rock-paper-scissors model, we will consider a QoI that provides a measure of how long each species has dominated the other. We will find the total time that P_r is near the carrying capacity of 1. The larger this is, the more dominant the species is. In Figure 10.2 we show the main effects from a factorial design.

10.3   State of the Art and Caveats

Factorial design is a method that is related to design of experiments. There are many other design methods that can be explored. Factorial design balances the number of simulations with robustness. In the other most used methods of assessing sensitivity, PRCC and Sobol', there is a question about how many samples to take. There is not conclusive theory. Factorial design avoids this question. This means that the time it takes to complete the simulations can be estimated with some precision. This can be quite useful when trying to develop a workflow that includes sensitivity analysis. This also means that using this may be inefficient for large numbers of parameters.

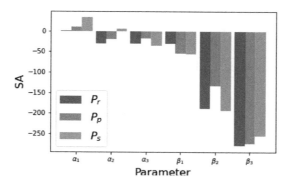

Figure 10.2: Factorial design sensitivity using a measure of dominance. We see relatively consistent results for each species.

Finally, because factorial design compares high and low values of parameters, it is important to have some understanding of the parameter impact – which is precisely what the method is trying to quantify. For example, taking too large an interval could miss important nonlinearities while taking too small an interval could lead to misclassification. This method is often used in conjunction with other methods to help support the conclusions.

10.4   Problems

Problems 10.1 *(a) Find the equilibria of the rock-paper-scissors model.*

(b) Use the numerical packages to solve the equations numerically.

(c) Vary the parameters and make a conjecture about how parameters alter the period of oscillation.

(d) Use the factorial design to confirm or disconfirm your conjecture.

Problems 10.2 *Use the framework of the rock-paper-scissors model,*

$$\frac{dp_r}{dt} = (a_r p_s - b_r p_p) p_r, \qquad (10.8)$$

$$\frac{dp_p}{dt} = (a_p p_r - b_p p_s) p_p, \qquad (10.9)$$

$$\frac{dp_s}{dt} = (a_s p_p - b_s p_r) p_s, \qquad (10.10)$$

to develop a model that has 10 compartments $p_1, ..., p_{10}$, with 20 parameters. Assume all parameters are in the interval $[0.9, 1.1]$. Compare the speed of the factorial design method for the case of three populations and ten populations.

Determine the total number of simulations that must be done for both the three and ten population examples.

Problems 10.3 *(a) Notice what happens if you take the numerical script for solving the rock-paper-scissor model and increase the final time. At some point the plot disappears. Diagnose what cause the plot to disappear (it is a much different question than explaining why the solution behaves this way).*

(b) Show that the change of variables, $P_{rl} = \ln P_r$, $P_{pl} = \ln P_p$. $P_{sl} = \ln P_s$ leads to a system of equations of the form,

$$\frac{dP_{rl}}{dt} = 1 - \alpha_1 e^{P_{rl}} - \beta_1 P_{pl},$$

$$\frac{dP_{pl}}{dt} = 1 - \alpha_2 e^{P_{pl}} - \beta_2 P_{sl},$$

$$\frac{dP_{sl}}{dt} = 1 - \alpha_3 e^{P_{sl}} - \beta_3 P_{rl}.$$

(c) Show that the numerical solution is well behaved for longer.

(d) How can we change the QoI to measure this?

Bibliography

[1] Rustom Antia, Jacob C Koella, and Veronique Perrot. Models of the within-host dynamics of persistent mycobacterial infections. *Proceedings of the Royal Society of London. Series B: Biological Sciences*, 263(1368):257–263, 1996.

[2] Nicolas Bacaër. Lotka, volterra and the predator–prey system (1920–1926). In *A Short History of Mathematical Population Dynamics*, pages 71–76. Springer, 2011.

[3] John C Baez. The fundamental theorem of natural selection. *Entropy*, 23(11):1436, 2021.

[4] Mary Ballyk, Don Jones, and HL Smith. The biofilm model of freter: a review. *Structured population models in biology and epidemiology*, pages 265–302, 2008.

[5] Jake Baum, Geoffrey Pasvol, and Richard Carter. The r. sub. 0 journey: from 1950s malaria to covid-19. *Nature*, 582(7813):488–489, 2020.

[6] Erin N Bodine, Suzanne Lenhart, and Louis J Gross. *Mathematics for the Life Sciences*. Princeton University Press, 2014.

[7] William E Boyce, Richard C DiPrima, and Douglas B Meade. *Elementary differential equations and boundary value problems*. John Wiley & Sons, 2021.

[8] Anna Bramwell. Ecology in the twentieth century: a history. 1990. Yale University Press.

[9] Nicholas Britton. *Essential Mathematical Biology*. Springer Science & Business Media, 2005.

[10] Stanca M Ciupe and Jane M Heffernan. In-host modeling. *Infectious Disease Modelling*, 2(2):188–202, 2017.

[11] Michael J Cooney, Jonathan Murry, Peter L Lee, and Michael R Johns. The importance of differential sensitivity analysis in the modelling, optimisation, and control of fermentation processes. *IFAC Proceedings Volumes*, 31(8):121–128, 1998.

[12] Athel Cornish-Bowden. One hundred years of michaelis–menten kinetics. *Perspectives in Science*, 4:3–9, 2015.

[13] Sarah C Cotter. A screening design for factorial experiments with interactions. *Biometrika*, 66(2):317–320, 1979.

[14] Leah Edelstein-Keshet. *Mathematical Models in Biology*. SIAM, 2005.

[15] Frank N Egerton. *Roots of ecology: antiquity to Haeckel*. Univ of California Press, 2012.

[16] Paul R Ehrlich, Jonathan Roughgarden, and Joan Roughgarden. *The science of ecology*. Number 574.5 E4. Macmillan New York, 1987.

[17] Ronald A Fisher. The genetical theory of natural selection. clarendon, 1930.

[18] Ronald A Fisher and Winifred A Mackenzie. Studies in crop variation. ii. the manurial response of different potato varieties. *The Journal of Agricultural Science*, 13(3):311–320, 1923.

[19] Michael Friendly and Daniel Denis. The early origins and development of the scatterplot. *Journal of the History of the Behavioral Sciences*, 41(2):103–130, 2005.

[20] George Francis Gause. *The struggle for existence*. Courier Corporation, 2003.

[21] Georgii Frantsevich Gause. Experimental studies on the struggle for existence: I. mixed population of two species of yeast. *Journal of experimental biology*, 9(4):389–402, 1932.

[22] John Graunt. *Natural and political observations made upon the bills of mortality*. Number 2. Johns Hopkins Press, 1939.

[23] Arthur C Guyton. Cardiac output and its regulation. *Circulatory physiology*, pages 353–371, 1973.

[24] David M Hamby. A review of techniques for parameter sensitivity analysis of environmental models. *Environmental monitoring and assessment*, 32(2):135–154, 1994.

[25] Godfrey Harold Hardy and Charles Percy Snow. *A Mathematician's Apology. Reprinted, with a Foreword by CP Snow*. Cambridge University Press, 1967.

[26] Johan Andre Peter Heesterbeek. A brief history of r 0 and a recipe for its calculation. *Acta biotheoretica*, 50(3):189–204, 2002.

[27] Jane M Heffernan, Robert J Smith, and Lindi M Wahl. Perspectives on the basic reproductive ratio. *Journal of the Royal Society Interface*, 2(4):281–293, 2005.

[28] Jon Herman and Will Usher. SALib: An open-source python library for sensitivity analysis. *The Journal of Open Source Software*, 2(9), jan 2017.

[29] Alan L Hodgkin and Andrew F Huxley. Action potentials recorded from inside a nerve fibre. *Nature*, 144(3651):710–711, 1939.

[30] Alan L Hodgkin and Andrew F Huxley. Resting and action potentials in single nerve fibres. *The Journal of physiology*, 104(2):176–195, 1945.

[31] AF Huxley. Hodgkin and the action potential 1935–1952. *The Journal of physiology*, 538(Pt 1):2, 2002.

[32] Sobol' IM. Sensitivity estimates for nonlinear mathematical models. *Math. Model. Comput. Exp*, 1(4):407–414, 1993.

[33] Ronald L Iman and William J Conover. The use of the rank transform in regression. *Technometrics*, 21(4):499–509, 1979.

[34] Bertrand Iooss and Paul Lemaître. A review on global sensitivity analysis methods. In *Uncertainty management in simulation-optimization of complex systems*, pages 101–122. Springer, 2015.

[35] Kenneth A Johnson and Roger S Goody. The original michaelis constant: Translation of the 1913 michaelis–menten paper. *Biochemistry*, 50(39):8264–8269, 2011.

[36] James P Keener and James Sneyd. *Mathematical Physiology*, volume 1. Springer, 1998.

[37] Maurice G Kendall. Partial rank correlation. *Biometrika*, 32(3/4):277–283, 1942.

[38] William Ogilvy Kermack and Anderson G McKendrick. A contribution to the mathematical theory of epidemics. *Proceedings of the royal society of london. Series A, Containing papers of a mathematical and physical character*, 115(772):700–721, 1927.

[39] Vladimir A Kuznetsov, Iliya A Makalkin, Mark A Taylor, and Alan S Perelson. Nonlinear dynamics of immunogenic tumors: parameter estimation and global bifurcation analysis. *Bulletin of mathematical biology*, 56(2):295–321, 1994.

[40] Joseph Lee Rodgers and W Alan Nicewander. Thirteen ways to look at the correlation coefficient. *The American Statistician*, 42(1):59–66, 1988.

[41] Chuanqi Li, Wei Wang, Jianzhi Xiong, and Pengyu Chen. Sensitivity analysis for urban drainage modeling using mutual information. *Entropy*, 16(11):5738–5752, 2014.

[42] Chia-Chiao Lin and Lee A Segel. *Mathematics Applied to Deterministic Problems in the Natural Sciences*. SIAM, 1988.

[43] Thomas Robert Malthus. *An Essay on the Principle of Population*. 1872.

[44] Simeone Marino and Denise E Kirschner. The human immune response to mycobacterium tuberculosis in lung and lymph node. *Journal of theoretical biology*, 227(4):463–486, 2004.

[45] Richard McGehee and Robert A Armstrong. Some mathematical problems concerning the ecological principle of competitive exclusion. *Journal of Differential Equations*, 23(1):30–52, 1977.

[46] Gregory J McRae, James W Tilden, and John H Seinfeld. Global sensitivity analysis—a computational implementation of the fourier amplitude sensitivity test (fast). *Computers & Chemical Engineering*, 6(1):15–25, 1982.

[47] Jacques Monod. The growth of bacterial cultures. *Annual review of microbiology*, 3(1):371–394, 1949.

[48] Max D Morris. Factorial sampling plans for preliminary computational experiments. *Technometrics*, 33(2):161–174, 1991.

[49] James D Murray. *Mathematical Biology II: Spatial Models and Biomedical Applications*, volume 3. Springer New York, 2001.

[50] James Dickson Murray. *Mathematical Biology I. An Introduction*. Springer, 2002.

[51] Denis Noble. Cardiac action and pacemaker potentials based on the hodgkin-huxley equations. *Nature*, 188(4749):495–497, 1960.

[52] Eugene P Odum. *Fundamentals of ecology*. WB Saunders company, 1959.

[53] Sarah P Otto and Troy Day. *A Biologist's Guide to Mathematical Modeling in Ecology and Evolution*. Princeton University Press, 2011.

[54] Karl Pearson. Vii. note on regression and inheritance in the case of two parents. *proceedings of the royal society of London*, 58(347-352):240–242, 1895.

[55] Punnett, Reginald Crundall. Early days of genetics. *Heredity*, 4:1–10, 1950.

[56] Andrea Saltelli. Sensitivity analysis for importance assessment. *Risk analysis*, 22(3):579–590, 2002.

[57] Andrea Saltelli, Paola Annoni, Ivano Azzini, Francesca Campolongo, Marco Ratto, and Stefano Tarantola. Variance based sensitivity analysis of model output. design and estimator for the total sensitivity index. *Computer physics communications*, 181(2):259–270, 2010.

[58] Martin Schuster, Eric Foxall, David Finch, Hal Smith, and Patrick De Leenheer. Tragedy of the commons in the chemostat. *PloS one*, 12(12):e0186119, 2017.

[59] Lee A Segel et al. *Mathematical Models in Molecular Cellular Biology*. CUP Archive, 1980.

[60] Barry Sinervo and Curt M Lively. The rock–paper–scissors game and the evolution of alternative male strategies. *Nature*, 380(6571):240–243, 1996.

[61] David L Smith, Katherine E Battle, Simon I Hay, Christopher M Barker, Thomas W Scott, and F Ellis McKenzie. Ross, macdonald, and a theory for the dynamics and control of mosquito-transmitted pathogens. *PLoS pathogens*, 8(4):e1002588, 2012.

[62] Charles Spearman. The proof and measurement of association between two things. 1961.

[63] Anne Talkington, Claudia Dantoin, and Rick Durrett. Ordinary differential equation models for adaptive immunotherapy. *Bulletin of mathematical biology*, 80(5):1059–1083, 2018.

[64] Zhike Zi. Sensitivity analysis approaches applied to systems biology models. *IET systems biology*, 5(6):336–346, 2011.

Index